DK SMITHSONIAN

TRAIN

THE DEFINITIVE VISUAL HISTORY

**LONDON, NEW YORK, MELBOURNE,
MUNICH, AND DELHI**

DORLING KINDERSLEY
Senior Editors Sam Atkinson, Jemima Dunne, Kathryn Hennessy
Senior Art Editor Sharon Spencer
Project Art Editor Amy Child
Editors Suhel Ahmed, Rod Green, Alison Sturgeon, Miezen van Zyl
Editorial Assistance Alexandra Beeden
Design Assistance Alex Lloyd
Photographer Gary Ombler
Picture Research Nic Dean
DK Picture Library Claire Bowers, Claire Cordier, Romaine Werblow
Jacket Designers Amy Child, Mark Cavanagh
Jacket Editor Maud Whitney
Jacket Design Development Manager Sophia MTT
Producer, Pre-Production Nikoleta Parasaki
Producer Linda Dare
Managing Editor Esther Ripley
Managing Art Editor Karen Self
Publisher Sarah Larter
Art Director Phil Ormerod
Associate Publishing Director Liz Wheeler
Publishing Director Jonathan Metcalf

DK INDIA
Managing Editors Pakshalika Jayaprakash, Rohan Sinha
Managing Art Editors Arunesh Talapatra, Sudakshina Basu
Senior Editor Anita Kahar
Senior Art Editors Chhaya Sajwan, Mahua Sharma
Project Editor Antara Moitra
Project Art Editor Vaibhav Rastogi
Editor Vibha Malhotra
Art Editors Namita, Supriya Mahajan, Divya PR, Devan Das
Assistant Art Editors Roshni Kapur, Vansh Kohli, Riti Sodhi
Production Manager Pankaj Sharma
Pre-production Manager Balwant Singh
Senior DTP Designers Sachin Singh, Jagtar Singh
DTP Designers Nand Kishor Acharya, Bimlesh Tiwary
Picture Researcher Aditya Katyal
Picture Research Manager Taiyaba Khatoon

General Consultant Tony Streeter
Contributors Julian Holland, Keith Fender
Gary Boyd-Hope, Jonathan Randle Falconer, Peter Herring, Keith Langston, Ashwani
Lohani, Malcolm McKay, David Wilcock

SMITHSONIAN
Smithsonian Curator L. Susan Tolbert, Deputy Chair and Curator,
Work and Industry Division, National Museum of American History,
Kenneth E. Behring Center, Smithsonian Institution
Smithsonian Enterprises
Product Development Manager, Licensing Kealy Wilson
Licensing Manager Ellen Nanney
Vice President, Education and Consumer Products Brigid Ferraro
Senior Vice President, Education and Consumer Products Carol LeBlanc

First American Edition, 2014

Published in the United States by
DK Publishing, 375 Hudson Street, New York, New York 10014

14 15 16 17 18   10 9 8 7 6 5 4 3 2 1

001—256473—Sept/2014

Published in Great Britain by Dorling Kindersley Limited.

A catalog record for this book is available from the Library of Congress.

ISBN 978-1-4654-2229-3

DK books are available at special discounts when purchased in bulk for sales
promotions, premiums, fund-raising, or educational use. For details, contact:
DK Publishing Special Markets, 375 Hudson Street, New York, New York 10014
or SpecialSales@dk.com.

Printed and bound in China by Leo Paper Products

Discover more at
www.dk.com

Contents

1804–1838: THE IRON HORSE

The invention of the steam-powered locomotive led to the
development of the first passenger railroad in Britain. This
new mode of transportation spread to other countries, with
Rocket setting the benchmark for future locomotives.

1839–1869: BUILDING NATIONS

New tracks were laid across Europe, the US, and India.
Meanwhile, engineers made further innovations to all
aspects of rail travel, increasing its speed and efficiency.
Mass transit began with the London Underground.

1870–1894:
A WORLD OF STEAM

The rapid growth of the railroads defined the power of human endeavor. Tracks negotiated every terrain and all kinds of obstacles, covering vast distances and making rail travel across continents possible. The glamour of rail travel was epitomized by grand stations and luxury services.

1895–1913: GOLDEN AGE

Electric-powered railroads rose to prominence in North America and Europe, while new innovations increased the efficiency of steam. Emulating London, Paris and New York introduced their own underground systems.

1914–1939: STEAM'S ZENITH

During World War I, locomotives were key in the transport of soldiers and munitions. After hostilities ended, steam trains became faster and streamlined, and diesel trains were rolled out for the first time in the US and Europe.

1940-1959: WAR AND PEACE

The destruction of many European rail lines during World War II and the redrawing of national borders at the end of the conflict forced many governments to overhaul their rail systems. Technological advances saw diesel and electric power take over from steam.

1960-1979: BUILT FOR SPEED

The Japanese "bullet" train heralded a new age of high-speed rail travel, inspiring Western countries to innovation on their own railroads. Increasing competition from road and air led to further modernization.

1980-1999: CHANGING TRACKS

New technology focused on developing high-speed networks throughout the world, but the period also saw the introduction of luxury trains. The Channel Tunnel opened, linking Britain to mainland Europe.

AFTER 2000: RAIL REVIVAL

The new millennium has seen China become a major proponent of rail travel, building tracks at an unprecedented rate and introducing new trains, including the ultrafast Maglev. On a global level, rail travel offered a more glamorous and luxurious alternative to the jetliner.

HOW RAILWAYS WORK: ENGINES AND TRACKS

This chapter offers an overview of basic rail technology, from how rails and locomotive wheels are designed, to signaling systems past and present. The engineering principles behind steam, diesel, and electric locomotives are explained.

The Rail Revolution

The click-clack of wheels on rails, the whiff of coal smoke and oil, a whistle in the distance, the feeling of anticipation and excitement at the start of a long journey...

Railroads capture our imagination. They speak to our soul. The elemental attractions of fire and steam, the fascination of technology, and the glamour of connecting faraway places have all helped cement the place of railroads in human hearts. For more than 200 years, trains have fueled ambitions and attracted groundbreaking engineers, inspiring them to create inventions that tapped into the human desire to move forward and open up a world of possibilities.

Most importantly, railroads have contributed to modern history in prosaic, practical ways. Arguably, no single tool has influenced today's industrial world more. From the first stuttering experiments in Cornwall and Wales in the United Kingdom, to the building of railroads that opened up whole continents and helped create nations, as they did in North America and elsewhere, to their capacity to make modern warfare feasible—the invention of the locomotive has shaped the globe, for good and bad.

Before the railroads, life moved at a different speed. Most people traveled only short distances from where they lived— there were no cars, no planes, no modern roads. Until the arrival of trains, there was no unified time and no compelling reason to introduce it. Towns and cities set their own time until the need for strict schedules on the railroads called for standardization. The new technology fueled urbanization— growing conurbations were fed by railroads, delivering people cheaply from ever farther afield. Rail networks moved commodities that previously could not be transported long distances—perishable fruit, newspapers, flowers, and fresh milk were delivered to the masses in a timely manner.

In these many ways, railroads became essential to the creation of modern life, and achieved it with panache. Companies gave their locomotives and services evocative names; they came up with attractive color schemes; and they worked hard on aesthetics to make their engines graceful, imposing, or dynamic, as well as functional. The drive to move ever forward shaped the railroads too. As new technologies developed, builders of new routes climbed higher, dug deeper, and went farther, taming the most inhospitable ground. The push to be ever faster, ever safer, and ever more efficient drove that progress too.

Across the globe, railroads put great effort into achieving higher speeds, into selling the luxury of their most exclusive trains, and into persuading people to use their services for both business and leisure. Modern marketing, public relations, the seaside vacation—in all these areas, the railroad has been an instrument of change and a driving force. It is no wonder that

> # "The **locomotive** is the **true harbinger of civilization**."
> BRITISH JOURNALIST AND EXPLORER HENRY MORTON STANLEY, MAY 9, 1867

schoolboys have dreamed of becoming locomotive engineers, that authors as diverse as Leo Tolstoy, Émile Zola, Agatha Christie, and Sir Arthur Conan Doyle have bound railroads into their dramas and mysteries, or that popular train-based songs like "Chattanooga Choo Choo" and "The Loco-Motion" have stood the test of time.

In the "Golden Age" of rail travel, newspapers and newsreels breathlessly reported the latest advances—as well as the gory details of crashes. New express engine designs were described in detail, engineers and designers became heroes, and there was fierce competition for headlines. Locomotives such as the huge "Big Boy" class in the US, or Britain's *Mallard*—which broke the speed record for steam in 1938 and still holds it—became famous the world over. Half a century after steam disappeared across large parts of the globe, it is still an emotive force—even among those who are too young to remember it in service. Dedicated enthusiasts chase the final survivors of the steam age in the most inaccessible places, or restore and preserve engines, coaches, and even entire lines.

It's not all been positive, and there's no denying the darker deeds that were made possible by railroads. They offered an opportunity for mass transit that enabled huge armies to be moved and supplied across continents, as well as the deportation of millions of people to Hitler's extermination camps during World War II. War became global and more deadly, and it was inevitable that rail networks would themselves become targets and face huge destruction in modern conflicts.

Yet while railroads entered increasingly difficult times—and after World War II, a period came when they were often seen as bland, monotonous, and outdated—they always resisted becoming merely a thing of the past. In recent years there's been a renaissance as countries have directed energy into building new high-speed routes and reducing their reliance on the automobile.

Today, long container trains are still a vital component in freight delivery, rumbling across continents as the pioneering trains did more than 200 years ago. Passengers speed across borders without having to leave their seats, and the idea of moving people quickly over long distances in comfort is once again in vogue. Technology forges ahead, the glamour has returned, and for many, the train is once again perceived as the civilized way to travel. Two centuries on from the pioneering moves of the first "iron horses," rail's exciting journey continues.

TONY STREETER
GENERAL CONSULTANT

1804-1838
THE IRON HORSE

THE IRON HORSE

In South Wales in February 1804, a new machine won a bet for its owner. The Pen-y-darren steam locomotive had just hauled wagons loaded with iron and people for nearly 10 miles (16 km). Richard Trevithick's machine was slow and cumbersome, but the achievement would soon change the world—the benefits of steam were felt quickly, and innovation was rapid.

In 1808, Trevithick's *Catch Me Who Can* pulled people around a short piece of circular track in London, but it was not until 1825 that passenger railroads really began to take off. In September of that year, the world's first passenger line to be paid for by public subscription was opened between Stockton and Darlington in northeast England. Its first locomotive was *Locomotion No. 1*, created by the father-and-son team of George and Robert Stephenson. The new technology rapidly went international; in France, engineer Marc Séguin built his own locomotive in 1828, and in 1829 the British-built *Stourbridge Lion* brought the steam age to the US, on the Delaware and Hudson line.

Britain's famous Rainhill Trials were held in 1829 to decide which locomotives would be built for the world's first intercity line, the Liverpool & Manchester Railway (1830). Triumph went to the Stephensons, with their *Rocket*. Many of the engine's innovations were so successful that *Rocket* set the basic layout for future locomotives right up until the end of the steam era. Engineers on both sides of the Atlantic began to adapt designs for their own terrain. In 1830 the US's home-built *Tom Thumb* made its debut on the Baltimore & Ohio Railroad.

△ **Stephensons' engines**
The *Rocket* was far from the Stephensons' only locomotive success; *North Star* ran on the Great Western Railway between 1838 and 1871.

> " The introduction of so **powerful an agent** as steam to a carriage on wheels will make **a great change** in the **situation of man.**"
THOMAS JEFFERSON, PRESIDENT OF THE UNITED STATES

◁ **Rainhill's skew arch bridge** was opened in 1830, one year after the famous Rainhill Trials

Key Events

▷ **1804** A locomotive at Pen-y-darren Colliery, South Wales, launches the steam age. An earlier design by Trevithick had apparently been built, but little information survives.

▷ **1808** Trevithick's *Catch Me Who Can* is demonstrated in London.

▷ **1813-14** William Hedley builds *Puffing Billy* to run at Wylam Colliery in northeast England.

▷ **1814** George Stephenson constructs his first locomotive, for Killingworth Colliery near Newcastle-upon-Tyne. It gains the name *Blücher*.

▷ **1825** The opening of Great Britain's Stockton & Darlington Railway, the world's first passenger railroad paid for by public subscription.

▷ **1828** Marc Séguin builds France's first locomotive.

▷ **1829** *Rocket*, the template for future locomotives, wins the Rainhill Trials.

▷ **1830** *Tom Thumb* and *The Best Friend of Charleston* herald the start of American locomotive building.

△ **Tom Thumb**
On the trial run of Peter Cooper's *Tom Thumb* in 1830 (reenacted here), the locomotive hauled a car containing 18 directors of the B&O Railroad.

▷ **1834** The first railroad in Ireland is opened between Dublin and Kingstown.

▷ **1835** Germany's first railroad opens between Nuremberg and Fürth.

▷ **1835** Great Britain's Great Western Railway is incorporated.

Richard Trevithick 1771–1833

Although Richard Trevithick built the world's first working steam locomotives, his name is less familiar than that of other pioneers in British railway engineering. Trevithick's misfortune was that he invented many of his machines 20 years before the world was ready to use them. As well as his locomotives, Trevithick pursued other engineering projects, which included a paddle-wheel barge, a steam hammer, a steam rolling mill, and a tunnel under the Thames River. Trevithick also spent 10 years in South America, where he used his steam engine designs to help open up silver mines in Peru. He returned to England in 1827, but died penniless six years later.

EVOLVING DESIGNS

Early on in his career, Trevithick worked as an engineer in the local mines in Cornwall. His familiarity with stationary engines used for winding and pulling meant that he was well placed to experiment with high-pressure engines or "strong steam," which offered greater pulling power for locomotives. He built his first steam vehicle, known as the *Puffer*, in 1801. This ran on roads, not rails, but met an unfortunate end when it crashed into a house in Camborne, Cornwall, and caught fire.

In 1803 Trevithick built a high-pressure engine that could run on iron rails. The Pen-y-Darren engine hauled 10 tons of iron and 70 passengers along a 9½-mile (15.3-km) iron rail track, proving its usefulness. Although the engine's weight eventually fractured the track, it was a milestone in the development of the locomotives.

Trevithick developed his third and final locomotive in 1808 and named it *Catch-me-who-can*. The train pulled passengers around a circular cast-iron track he had built in London. Eventually the weight of the train fractured the track and derailed the engine, but by then Trevithick had proven to the world that a steam locomotive could be run on tracks.

The money train
Catch-me-who-can (1809) was the first passenger train to charge a fare. People had to pay one shilling to ride the train, which traveled at just over 12 mph (19 km/h) around a circular demonstration track in London.

TRAM ENGINE
4¾ Inches DIA.ᵣ of Cylinder
3 Feet
Stroke
Decᵣ 1803

Running on rails
Trevithick's high-pressure tram engine (1803) was the world's first railroad locomotive, and a forerunner to the Stephensons's engines. It was used at the Pen-y-Darren ironworks in South Wales, where it ran successfully on rails.

A British Invention

During the 18th century, the British inventors Thomas Newcomen and James Watt led the way in the development of the low-pressure, stationary steam engines that played a vital role in the early years of the Industrial Revolution. A breakthrough took place in the early 19th century when Cornish inventor Richard Trevithick successfully demonstrated the world's first working, high-pressure steam locomotive. From then on, British inventiveness, led by the "Father of British Railways," George Stephenson, brought rapid development, which culminated in 1830 with the opening of the world's first intercity railroad, between Liverpool and Manchester.

△ **Pen-y-darren locomotive, 1804**

Wheel arrangement	0-4-0
Cylinders	1
Boiler pressure	25 psi (1.75 kg/sq cm)
Drive wheel diameter	48 in (1,220 mm)
Top speed	approx. 5 mph (8 km/h)

Richard Trevithick's high-pressure steam locomotive hauled the world's first train on the Pen-y-darren Ironworks tramway in Merthyr Tydfil, South Wales on February 13, 1804. The train carried 11.24 tons (10.2 metric tons) of coal and 70 men.

◁ *Catch Me Who Can*, **1808**

Wheel arrangement	2-2-0
Cylinders	1
Boiler pressure	25 psi (1.75 kg/sq cm)
Drive wheel diameter	48 in (1,220 mm)
Top speed	approx. 12 mph (19 km/h)

Richard Trevithick's *Catch Me Who Can* was demonstrated on a circular track at a steam circus in Bloomsbury, London, in 1808. Unfortunately, the train overturned when a rail broke, so the public was not convinced.

◁ *Puffing Billy*, **1813**

Wheel arrangement	0-8-0 (final form 0-4-0)
Cylinders	2
Boiler pressure	40 psi (2.8 kg/sq cm)
Drive wheel diameter	48 in (1,220 mm)
Top speed	approx. 5 mph (8 km/h)

Weighing 8.8 tons (8 metric tons) and built by William Hedley for the Wylam Colliery in Northumberland, *Puffing Billy* was the world's first commercial traction steam engine. Now preserved at London's Science Museum, it is considered the oldest surviving locomotive.

▷ *Locomotion No. 1*, **1825**

Wheel arrangement	0-4-0
Cylinders	2
Boiler pressure	50 psi (3.51 kg/sq cm)
Drive wheel diameter	48 in (1,220 mm)
Top speed	approx. 15 mph (24 km/h)

Built by George and Robert Stephenson, *Locomotion No. 1* hauled the first train on the Stockton & Darlington Railway, the world's first public railroad, in 1825. This locomotive has been preserved and can be seen at the Darlington Railway Museum, County Durham.

△ *Rocket*, 1829

Wheel arrangement 0-2-2
Cylinders 2
Boiler pressure 50 psi (3.51 kg/sq cm)
Drive wheel diameter 56³/₄ in
(1,435 mm)
Top speed approx. 30 mph (48 km/h)

Robert Stephenson & Co.'s advanced and
innovative *Rocket* was the clear winner of
the Rainhill Trials held on the Liverpool &
Manchester Railway in 1829. The *Rocket*
is shown pulling a first-class passenger
car; luggage was carried on the roof.

▷ *Agenoria*, 1829

Wheel arrangement 0-4-0
Cylinders 2
Boiler pressure 40 psi (2.8 kg/sq cm)
Drive wheel diameter 48 in (1,220 mm)
Top speed approx. 8 mph (13 km/h)

One of only four steam locomotives built
by Foster, Rastrick & Co. of Stourbridge,
Agenoria worked on the Earl of Dudley's
Shutt End Colliery Railway, Staffordshire,
for 35 years. The same company built the
Stourbridge Lion, the first locomotive to
be exported to the US.

◁ *Sans Pareil*, 1829

Wheel arrangement 0-4-0
Cylinders 2
Boiler pressure 50 psi (3.51 kg/sq cm)
Drive wheel diameter 54 in (1,372 mm)
Top speed approx. 18 mph (29 km/h)

Built by Timothy Hackworth, *Sans Pareil*
(meaning "without equal") performed well
in the Rainhill Trials on the Liverpool &
Manchester Railway in 1829 but exceeded
the permitted weight, so it was not
considered for the prize.

▷ *Novelty*, 1829

Wheel arrangement 0-2-2WT
Cylinders 2
Boiler pressure 50 psi (3.51 kg/sq cm)
Drive wheel diameter 54 in (1,372 mm)
Top speed approx. 28 mph (45 km/h)

Although it was one of the fastest
locomotives at the 1829 Rainhill Trials,
John Ericsson and John Braithwaite's
lightweight *Novelty* proved unreliable and
was withdrawn. It was the first locomotive
to have its cylinders within the frames.

Rocket

The Rainhill Trials were staged in 1829 to decide which locomotives would run the world's first "intercity" passenger trains on the Liverpool & Manchester Railway (L&MR) from September 15, 1830. Built by engineer Robert Stephenson, *Rocket* competed in the trials and hit a top speed of 28 mph (45 km/h). As the undisputed winner, *Rocket* clinched the prized contract, winning fame and universal acclaim for Stephenson.

ROCKET FEATURED A NUMBER of engineering innovations that ensured its success at the Rainhill Trials. It had inclined cylinders on either side of the firebox, which were connected to single drive wheels by short rods, giving it more thrust than could be achieved by the beam arrangement on earlier engines. It was the first engine to have a multitube boiler and stack blastpipe, which greatly improved steam production. The basic design principles embodied in *Rocket* carried through to the last steam locomotives. The original 1829 *Rocket* can be seen in London's Science Museum, but was extensively modified. The replica shown here is a more accurate representation of the original build. A working replica built in 1979 for the 150th anniversary of the L&MR resides at the National Railway Museum in York. It incorporates the trailing and tender wheelsets and iron frame from a replica built at Crewe Works in 1880 to mark the centenary of George Stephenson's birth.

SPECIFICATIONS FOR ORIGINAL ROCKET	
Class	Rocket
Wheel arrangement	0-2-2
Origin	UK
Designer/builder	Robert Stephenson & Co.
Number produced	5
In-service period	1830-40
Cylinders	2, inclined at 37 degrees
Boiler pressure	50 psi (3.51 kg/sq cm)
Drive wheel diameter	56¾ in (1,435 mm)
Top speed	approx. 30 mph (48 km/h)

Stack coronet

16-ft (5-m) stack

Stack stays

Inclined cylinders

Dome

Water barrel

Fuel space

CROSS-SECTION OF *ROCKET* AND TENDER (ABOVE)   FIRST-CLASS CARS (BELOW)

Roof luggage rack

Six-seat compartment

Hand-painted company name

Sprung buffers with safety chain link

Car name

Oak-framed windows

Conductor's seat

Revolutionary engine
This *Rocket* replica includes many of the features that facilitated the speeds the original achieved at the trials. *Rocket* was the first locomotive to have a fully functional blastpipe, which forced exhaust steam up the stack. The engine had no brakes. Stopping was only achieved via a foot pedal that puts the engine into reverse gear.

SIDE VIEW

FRONT VIEW

ROCKET.

EXTERIOR

The 1979 working replica *Rocket* could never be wholly faithful to the original because of the need to conform to modern health and safety requirements and operating conditions. The replica was initially equipped with wooden drive wheels, which had to be replaced with steel ones when the originals buckled as the engine derailed on rough track at Bold Colliery in May 1980, forcing it to miss the opening of the *Rocket 150* celebrations. The replica's "water tank" was a 65-gallon (245-liter) "Hogshead" beer barrel originally from the Wadworth Brewery in Devizes, England.

1. Boiler barrel brass nameplate **2.** Stack coronet **3.** Pressure gauge shut-off valve
4. Boiler water-level test taps **5.** Crosshead, showing connecting rod small end and piston rod
6. Wooden drive wheel and connecting rod **7.** Small end detail **8.** Works plate **9.** Left side
driving rod **10.** Slip eccentric assembly **11.** Water barrel **12.** Tender wheel and safety chains

FOOTPLATE

Rocket has a small basic footplate that provides no weather protection for the crew. The "fallplate" (the metal plate that bridges the gap between engine and tender) rocks and slides around when in operation, so in wet and windy weather the engineer can feel as if he is at sea. The propensity for dropped lumps of coke or coal to lodge beneath the floor-mounted, valve-gear treadle and firebox-damper handle could make driving conditions particularly difficult.

13. Firebox with copper, main-steam feed pipes above   **14.** Valve gear operating treadle   **15.** Regulator valve   **16.** Firebox door   **17.** Right-side, valve-gear control levers

CARS

The cars seen with the *Rocket* replica are reproductions of original 1834 L&MR first-class coaches, built in 1930 for the railroad's centennial. The cars each have three six-seat compartments, and are named *Traveller* and *Huskisson*—the latter after Liverpool Member of Parliament William Huskisson, who was struck and killed by *Rocket* at the L&MR's opening ceremony in 1830.

18. Car buffer   **19.** Tender buffer spring **20.** Brass hand grip   **21.** Car name in gold leaf **22.** Car steps for passengers   **23.** Car wheel, axle box, and leaf spring   **24.** Guard's seat   **25.** Car window strap   **26.** First-class "button-back" upholstered seats

The Liverpool & Manchester Railway

The Liverpool & Manchester Railway (L&MR) opened in 1830 and was the world's first railroad to carry both fare-paying passengers and freight. It established a cheaper and more efficient transportation link between the factories of Manchester and Lancashire, England, and the port of Liverpool. This delighted factory owners, who sought faster and cheaper routes than those provided by boats on the Bridgewater Canal.

ENGINEERING CHALLENGE

George Stephenson was appointed engineer of the twin-track, 32-mile (51.5-km) line. His expertise was put to the test at Chat Moss near Manchester, where the track crossed an unstable 4-mile (6.5-km) stretch of peat bog. The terrain almost brought the project to a halt, but Stephenson overcame the challenge by having the line built on a floating foundation of wood and stone.

Within three years, 64 bridges and viaducts had been constructed along the line, including the nine-arch Sankey Viaduct, the Wapping Tunnel in Liverpool, and a 2-mile (3.2-km) cutting through Olive Mount. A passenger terminus was also built at each end of the line in Manchester and Liverpool. In the first six months of 1831, the L&MR carried 188,726 passengers and almost 36,000 tons (36,578 metric tons) of freight.

From top to bottom: *Liverpool* with first-class cars and a mail car; *Fury* with second-class cars; *North Star* pulling boxcars; and *Jupiter* transporting livestock.

Steam for Home and Export

The success of the Stephensons's *Rocket* and the opening of the world's first public railway in 1825 and the inter-city route in 1830 led to demand for British-built steam railroad locomotives at home and abroad. The most successful of the early builders was Robert Stephenson & Company of Newcastle, founded by George and his son Robert in 1823. Its early locomotives were built for the Stockton & Darlington Railway but it also supplied locomotives for the first railroads in Egypt and Germany as well as the US.

◁ *Invicta*, **1829-30**

Wheel arrangement	0-4-0
Cylinders	2
Boiler pressure	40 psi (2.81 kg/sq cm)
Driving wheel diameter	48 in (1,220 mm)
Top speed	approx. 20 mph (32 km/h)

Robert Stephenson & Co. built *Invicta* in Newcastle, then shipped it to Kent (UK) by sea. *Invicta* hauled the first train on the Canterbury & Whitstable Railway in 1830. The locomotive was named after the motto "invicta" (undefeated) on the flag of Kent. It is on display at Kent's Canterbury Museum.

▽ *John Bull*, **1831**

Wheel arrangement	0-4-0 (as built)
	2-4-0 (as modified)
Cylinders	2 (inside)
Boiler pressure	45 psi (3.16 kg/sq cm)
Driving wheel diameter	66 in (1,676 mm)
Top speed	approx. 30 mph (48 km/h)

Built by Robert Stephenson & Co., *John Bull* was exported to the US, where it worked on the Camden & Amboy Railroad from 1831 to 1866. US engineer Isaac Dripps added his two-wheel bogie, to which he attached the first cowcatcher, as well as a headlight, spark-arresting chimney, and covered tender and cab.

▷ *Planet*, **1830**

Wheel arrangement	2-2-0
Cylinders	2 (inside)
Boiler pressure	45 psi (3.16 kg/sq cm)
Driving wheel diameter	66 in (1,676 mm)
Top speed	approx. 35 mph (56 km/h)

Planet was the first type to have inside cylinders and the ninth locomotive built for the Liverpool & Manchester Railway. Designed by Robert Stephenson & Co., *Planet* was the first engine type to be built in large numbers.

JOHN BULL AS FIRST CONSTRUCTED, 1831

▽ *Adler*, 1835

Wheel arrangement	2-2-2
Cylinders	2 (inside)
Boiler pressure	48 psi (3.37 kg/sq cm)
Driving wheel diameter	54 in (1,372 mm)
Top speed	approx. 17 mph (27 km/h)

The *Adler* (meaning "eagle") was the first successful steam railroad locomotive to operate in Germany. It was built for the Bavarian Ludwig Railway by Robert Stephenson & Co. *Adler* remained in service until 1857. In 1935 a replica was built to mark the centenary of the German railroads.

▽ **Bury**, 1831

Wheel arrangement	0-4-0
Cylinders	2 (inside)
Boiler pressure	50 psi (3.52 kg/sq cm)
Driving wheel diameter	66 in (1,676 mm)
Top speed	approx. 40 mph (64 km/h)

These locomotives were built with bar frames to reduce weight and were noted for their round-topped fireboxes. Designed by Edward Bury & Co., the Bury was popular in the US where light track was laid quickly to cover vast distances.

△ Hawthorn *Sunbeam*, 1837

Wheel arrangement	2-2-0
Cylinders	2 (inside)
Boiler pressure	50 psi (3.52 kg/sq cm)
Driving wheel diameter	60 in (1,524 mm)
Top speed	approx. 40 mph (64 km/h)

Sunbeam was built by R. & W. Hawthorn & Co. of Newcastle for the Stockton & Darlington Railway. Hawthorn built marine and stationary steam engines as well as locomotives for the broad-gauge Great Western Railway.

◁ *North Star*, 1838

Wheel arrangement	2-2-2
Cylinders	2 (inside)
Boiler pressure	50 psi (3.52 kg/sq cm)
Driving wheel diameter	84 in (2,134 mm)
Top speed	approx. 40 mph (64 km/h)

Robert Stephenson & Co.'s *North Star* hauled the inaugural director's train on the broad-gauge Great Western Railway in 1838. The locomotive was rebuilt in 1854 and withdrawn from service in 1871.

▷ *Lion*, 1838

Wheel arrangement	0-4-2
Cylinders	2 (inside)
Boiler pressure	50 psi (3.52 kg/sq cm)
Driving wheel diameter	60 in (1,520 mm)
Top speed	approx. 35 mph (56 km/h)

Lion was one of the first two locomotives built by Todd, Kitson & Laird. The other one was called *Tiger*. *Lion* worked on the Liverpool & Manchester Railway until 1859 before it was retired to Liverpool Docks as a stationary pumping engine.

The Stephensons
1781–1848/1803–59

GEORGE STEPHENSON
1781–1848

ROBERT STEPHENSON
1803–59

In 1830, the opening of the world's first passenger railroad, the Liverpool & Manchester, heralded the dawn of mechanized transportation. The man responsible was George Stephenson, a self-taught colliery engineer, who is known as the "Father of the Railways" for his pioneering achievements in civil and mechanical engineering. Working with his engineer son Robert, Stephenson created a series of steam locomotives. The pair also collaborated on building the Stockton & Darlington Railway (1825), where George introduced his standard 4 ft 8½ in (1.44 m) rail gauge, which is still in use worldwide today.

A GROWING REPUTATION

George Stephenson was an innovator from the start. In 1814 he built his inaugural locomotive, *Blücher*, which was the first engine to use flanged wheels running on rails. In 1823 he set up the locomotive works in Newcastle, England, with Robert that built the first steam engines to run on commercial train lines. The company's first engine was named *Locomotion No. 1*, but perhaps the best known was *Rocket*, which serviced Liverpool and Manchester Railway after winning a competition in 1829.

The Stephensons' growing reputation meant that they were much in demand as chief engineers to Great Britain's burgeoning rail network, following the Liverpool & Manchester with the London & Birmingham railroad in 1833. They were even consulted on railroad plans overseas, in Egypt, Italy, and Norway. Robert's expertise also extended to rail bridges; he engineered the High Level Bridge in Newcastle (1849) and the Royal Border Bridge in Northumberland (1850), among others.

Digging deep
Approximately 480,000 cubic yards (367,000 cubic meters) of rock were excavated for the 2-mile- (3.2-km-) long Olive Mount Cutting on the Liverpool & Manchester Railway. The cutting is almost 70 ft (21 m) deep in places.

Winner takes all
Steam locomotive trials were run in October 1829 at Rainhill, near Liverpool, to decide which engine would be used on the Liverpool & Manchester Railway. Stephenson's *Rocket* triumphed, beating four other competitors.

World Pioneers

By the mid-1820s, pioneering inventors in continental Europe and the United States were experimenting with their own designs. Some of these developments, such as US civil engineer John B. Jervis's swiveling leading truck or Frenchman Marc Séguin's multitube boiler, would soon be incorporated into locomotives around the world. By the late 1830s, rapid technological advances in steam locomotive design led to a massive expansion of railroad building. In the US, the Baltimore & Ohio Railroad was the first to operate scheduled freight and passenger services. By 1837 the service had extended from Baltimore over the iconic Thomas Viaduct to Washington and across the Potomac River to Harper's Ferry.

▽ **John Stevens's *Steam Waggon*, 1825**

Wheel arrangement	early rack-and-pinion
Cylinders	1
Boiler pressure	approx. 100+ psi (7.03 kg/sq cm)
Drive wheel diameter	57 in (1,448 mm)
Top speed	approx. 12 mph (19 km/h)

Colonel John Stevens's *Steam Waggon* demonstrated the practicability of very-high-pressure steam railroad locomotives. This was the first engine to run on rails in the US. Stevens ran it on a circular track on his estate in Hoboken, New Jersey.

△ **Marc Séguin's locomotive, 1829**

Wheel arrangement	0-4-0
Cylinders	2
Boiler pressure	approx. 35 psi (2.46 kg/sq cm)
Drive wheel diameter	approx. 54 in (1,372 mm)
Top speed	approx. 15 mph (24 km/h)

Fitted with a multitube boiler, enormous rotary blowers, and a large firebox, Marc Séguin's innovative steam locomotive was the first to be built in France. It was tested on the Saint-Étienne & Lyon Railway in November 1829 and entered regular service in 1830.

The Best Friend, from William H. Brown, *The History of the First Locomotives in America*, New York, 1874

THE CHARLESTON & HAMBURG RAILROAD

△ ***Best Friend of Charleston*, 1830**

Wheel arrangement	0-4-0
Cylinders	2
Boiler pressure	approx. 35 psi (2.46 kg/sq cm)
Drive wheel diameter	approx. 57 in (1,448 mm)
Top speed	approx. 25 mph (40 km/h)

The first steam locomotive to be constructed entirely in the US, *Best Friend of Charleston* was built by the West Point Foundry in New York. It operated a passenger service on the South Carolina Railroad until it was destroyed by a boiler explosion.

JOHN STEVENS
1825

PIONEERS

Marc Séguin, 1786–1875

Born in the Ardèche region of France, engineer, inventor, and entrepreneur Marc Séguin built innovative steam locomotives for the Saint-Étienne & Lyon Railway. His engines were fitted with an ingenious multitube boiler, which he patented in 1827, as well as mechanically driven fans to improve drafting for the fire and a firebox enclosed by a water jacket for greater heating capacity. Séguin developed the first suspension bridge in continental Europe and went on to build 186 bridges in France.

Engineering Innovation Marc Séguin was inspired by George Stephenson's *Locomotion No.1*, which he saw in action on the Stockton & Darlington Railway in 1825.

◁ *Tom Thumb*, 1830

Wheel arrangement	2-2-0
Cylinders	1
Boiler pressure	approx. 35 psi (2.46 kg/sq cm)
Drive wheel diameter	approx. 33 in (840 mm)
Top speed	14 mph (23 km/h)

This locomotive was built by US merchant and inventor Peter Cooper. The Baltimore & Ohio Railroad raced *Tom Thumb* against a horse to decide whether they should adopt steam power or horse traction; the train lost, but the railroad saw its potential. Weighing only a ton, *Tom Thumb* had a vertical boiler with inner tubes fashioned from gun barrels.

▷ *DeWitt Clinton*, 1831

Wheel arrangement	0-4-0
Cylinders	2
Boiler pressure	approx. 35 psi (2.46 kg/sq cm)
Drive wheel diameter	approx. 58¾ in (1,520 mm)
Top speed	approx. 20 mph (32 km/h)

The first steam locomotive to operate in New York State, the *DeWitt Clinton* was built for the Mohawk & Hudson Railroad. Passengers traveled in converted stagecoaches. The locomotive was named after the governor of New York who was responsible for the construction of the Erie Canal.

◁ *Experiment*, 1832

Wheel arrangement	4-2-0
Cylinders	2
Boiler pressure	approx. 50 psi (3.51 kg/sq cm)
Drive wheel diameter	approx. 72 in (1,830 mm)
Top speed	approx. 60 mph (96 km/h)

This engine was designed by John B. Jervis, chief engineer for the Delaware & Hudson Canal & Railroad. *Experiment*, later named *Brother Johnathan*, was built by the West Point Foundry, New York, for use on the Mohawk & Hudson Railroad. It was the first locomotive with a leading truck that became the 4-2-0 type.

Railroad Expansion

The earliest US railroads were operated using horse power. In 1830 the Baltimore & Ohio Railroad (B&O) was one of the first to introduce steam. While some railroads bought designs from fledging manufacturers such as the Baldwin Locomotive Works of Philadelphia, the B&O started constructing their own. In 1836, William Norris introduced the four-wheel leading truck, which became common worldwide until the end of steam in the 20th century. Two years later, Johann Schubert's *Saxonia* became the first successful steam engine to be built and operated in Germany.

△ **Baldwin *Old Ironsides*, 1832**

Wheel arrangement	2-2-0
Cylinders	2
Boiler pressure	50 psi (3.51 kg/sq cm)
Drive wheel diameter	54 in (1,372 mm)
Top speed	approx. 28 mph (45 km/h)

Designed by US inventor Matthias Baldwin, *Old Ironsides* was the first commissioned steam locomotive. It was built by the Baldwin Locomotive Works for the Philadelphia, Germantown & Norristown Railroad.

▷ **B&O *Atlantic*, 1832**

Wheel arrangement	0-4-0
Cylinders	2
Boiler pressure	50 psi (3.52 kg/sq cm)
Drive wheel diameter	35 in (890 mm)
Top speed	approx. 20 mph (32 km/h)

Built by US inventor and foundry owner Phineas Davis for the Baltimore & Ohio Railroad, *Atlantic* was the prototype for 20 more similar locomotives nicknamed "Grasshoppers."

Early Coaches

The first railroad passenger coaches in the US were primitive affairs, often based on existing designs for turnpike stagecoaches and originally intended for low-speed, horse-operated railroads. The rail companies soon learned that they were impractical: seats were uncomfortable, and passengers in open-air cars had to brave not only the elements but also the smoke, hot ash, and cinders blown out by the equally primitive steam locomotives.

◁ **Director's Car, 1828**

Type	4-wheel
Capacity	12 passengers
Construction	iron and wood
Railway	Baltimore & Ohio Railroad

Originally horse-drawn, in August 1830 the Baltimore & Ohio Director's Car carried the railroad's directors in the first steam-hauled train along the railroad to Ellicott's Mills behind *Tom Thumb*. This is a replica built in 1926 for the Fair of the Iron Horse and can be seen at the B&O Railroad Museum, Baltimore, MD.

▷ B&O *Lafayette*, 1837

Wheel arrangement	4-2-0
Cylinders	2
Boiler pressure	60 psi (4.21 kg/sq cm)
Drive wheel diameter	54 in (1,372 mm)
Top speed	approx. 35 mph (56 km/h)

William Norris's *Lafayette* was the first in the world to feature a leading four-wheel truck on a production model. A replica, built in 1927, can be seen at the Baltimore & Ohio Railroad Museum, Baltimore, MD.

△ B&O "Grasshopper" *John Hancock*, 1836

Wheel arrangement	0-4-0
Cylinders	2
Boiler pressure	50 psi (3.51 kg/sq cm)
Drive wheel diameter	35 in (890 mm)
Top speed	approx. 20 mph (32 km/h)

Equipped with an engineer's cab, *John Hancock* was one of 20 "Grasshopper" locomotives built by the Baltimore & Ohio Railroad. It was in service as a switcher until 1892.

▷ *Saxonia*, 1838

Wheel arrangement	0-4-2
Cylinders	2
Boiler pressure	60 psi (4.21 kg/sq cm)
Drive wheel diameter	59 in (1,500 mm)
Top speed	approx. 37 mph (60 km/h)

Designed by Johann Schubert, *Saxonia* was the first practical working steam locomotive built entirely in Germany. It worked on the Leipzig to Dresden Railway, Germany's first long-distance line. By 1843, *Saxonia* had clocked up more than 5,300 miles (8,500 km).

◁ Maryland Coach, 1830

Type	4-wheel
Capacity	14 passengers
Construction	iron and wood
Railway	Baltimore & Ohio Railroad

Based on a stagecoach, Richard Imlay's double-deck coach was one of six built for the inaugural steam train on the Baltimore & Ohio Railroad. The car body was perched on unsprung wheels cradled on leather straps. It was unstable and offered little protection for top deck passengers.

▷ Nova Scotia Coach, 1838

Type	4-wheel
Capacity	6 passengers
Construction	iron and wood
Railway	General Mining Association of Nova Scotia

Built by Timothy Hackworth of London (UK), the Nova Scotia Coach carried the Director of Nova Scotia's General Mining Association on the colliery railroad on Cape Breton Island in Canada. Also known as the bride's car, it was said to have originally carried the director's new bride to their home after their marriage ceremony.

1839-1869
BUILDING NATIONS

BUILDING NATIONS

On May 10, 1869 a "**golden spike**" was hammered into a railroad tie in dusty Promontory, UT, and two locomotives eased gently toward each other. The simple ceremony marked the completion of the first transcontinental railroad and was a key moment in the development of the United States. In the mid-19th century, railroads were to become the driving force of progress not just in the US, but also throughout the world. Tracks spread across Europe and into ever more inaccessible places. In India, a country that would become one of the greatest railroad nations, the first passenger train left Bombay in April 1853.

△ **India's first passenger train**
On April 16, 1853, GIP No.1 carried passengers from Bombay (now Mumbai) to Tanneh (now Thane) on the Great Indian Peninsular Railway.

Yet this was still a time of experiments. Engineering genius Isambard Kingdom Brunel built Great Britain's Great Western Railway (GWR) not to the normal 4 ft 8 1/2 in (1.435 m) track gauge, but to his own much wider 7 ft 1/4 in (2.14 m). The bigger gauge allowed for high speeds and more spacious trains, but too much track had already been laid to the narrower width favored by Stephenson. Mediating in the "gauge war," the UK Parliament decided against Brunel's idea.

Other inventions had long-lasting effects: the telegraph and "mechanical interlocking" that connected signals and points were developed through the 1850s, and in 1869 George Westinghouse introduced air brakes, now standard around the world. As railroads grew, so did their reach through society; in 1842, Queen Victoria took her first train journey, traveling via the GWR on her way to Windsor. A fundamental change came with the birth of mass transit when London's first underground line opened in 1863. As the network developed, laborers could travel cheaply to work from the city's outskirts, fueling the creation of the world's first metropolis.

"Let the country **but make the railroads, and the railroads will make the country.**"
EDWARD PEASE, BRITISH RAILROAD PROMOTER

Key Events

▷ **1839** Germany's first long-distance line opens, linking the cities of Leipzig and Dresden.

▷ **1839** In the Netherlands, Amsterdam and Haarlem are connected by the country's first railroad.

▷ **1841** Thomas Cook invents the "charter" train, for temperance campaigners traveling from Leicester to Loughborough, England.

▷ **1842** Great Britain's Queen Victoria gives royal approval by traveling on the Great Western Railway.

▷ **1844** The railroad reaches Basel, Switzerland, via France; Switzerland's first domestic line opens in 1847.

▷ **1850s** Safety is improved with mechanical interlocking that connects points and signals together.

▷ **1853** India's debut passenger train runs from Bombay to Thane.

▷ **1863** The world's first true underground railroad—London's Metropolitan Railway—is opened.

△ **The Metropolitan Line**
Steam locomotives in tunnels meant that passengers had to contend with smoky stations and cars that were lit by gas lamps.

▷ **1869** North America's first transcontinental railroad is completed.

▷ **1869** George Westinghouse (US) invents the air brake.

▷ **1869** George Mortimer Pullman launches the ultimate in luxury travel—the Pullman Car.

◁ **The "golden spike," placed at Promontory, UT, in 1869,** linked the Union Pacific and Central Pacific railroads

The US Forges Ahead

The British locomotives imported into the United States were often too heavy for the lighter, quickly laid rail tracks, and not powerful enough to cope with the steeper gradients. So US engineers developed designs tailored to their railroad's needs. A leading truck, first with two wheels, then four, was installed to guide engines through sharp curves. Improved traction led to the 4-4-0 becoming the standard type, soon followed by the more powerful 4-6-0. Cowcatchers, headlights, and warning bells were added to cope with the unfenced tracks. American designers built locomotives capable of hauling heavy loads over a railroad system that by 1871 linked two oceans.

▷ B&O L Class No. 57 *Memnon*, 1848

Wheel arrangement	0-8-0
Cylinders	2
Boiler pressure	75 psi (5 kg/sq cm)
Drive wheel diameter	44 in (1,118 mm)
Top speed	approx. 30 mph (48 km/h)

Bought by the Baltimore & Ohio Railroad in 1848 for freight working, *Memnon* was later used in the Civil War, for hauling troops and supplies. Eight drive wheels gave this locomotive its extra power and traction.

▽ CVR No. 13 *Pioneer*, 1851

Wheel arrangement	2-2-2
Cylinders	2
Boiler pressure	100 psi (7 kg/sq cm)
Drive wheel diameter	54 in (1,372 mm)
Top speed	approx. 40 mph (64 km/h)

Pioneer hauled the short passenger trains of the Cumberland Valley Road of Pennsylvania and western Maryland until 1890. It survived the destruction of the railroad's workshops by Confederate troops in 1862.

Financing the Railroads

Railroad promoters looked to the commercial centers of Philadelphia, Boston, and New York, as well as European money markets, to raise capital to develop the railroads. Investors preferred bonds to stocks, since these offered a guaranteed income. At the same time, the US government offered federal land grants to the rail companies, who then sold the land they did not need to raise more funds.

B&O stocks The value of shares in the new Baltimore & Ohio Railroad exceeded $3 million in 1839 when this $100 certificate was issued.

▷ W&A No. 39 *The General*, 1855

Wheel arrangement	4-4-0
Cylinders	2
Boiler pressure	140 psi (10 kg/sq cm)
Drive wheel diameter	60 in (1,524 mm)
Top speed	approx. 45 mph (72 km/h)

Built by the Western & Atlantic Railroad, *The General* pulled passenger and freight trains between Atlanta, Georgia, and Chattanooga, Tennessee, from 1856 until 1891.

△ **B&O Class B No.147**
Thatcher Perkins, 1863

Wheel arrangement 4-6-0

Cylinders 2

Boiler pressure 175 psi (12.30 kg/sq cm)

Drive wheel diameter 60 in (1,524 mm)

Top speed approx. 50 mph (80 km/h)

The Baltimore & Ohio's *Thatcher Perkins* (named after the company's Master of Machinery who designed it) is a survivor from among 16,500 "Ten-Wheelers" (4-6-0s) that were built for American railroads up to 1910. Its power was deployed climbing the steeply graded lines of West Virginia.

▽ **UP No. 119, 1868**

Wheel arrangement 4-4-0

Cylinders 2

Boiler pressure 85 psi (6 kg/sq cm)

Drive wheel diameter 60 in (1,524 mm)

Top speed approx. 45 mph (72 km/h)

This is a replica of the Union Pacific's No. 119, first built by Roger's Locomotive Works of Paterson, New Jersey. The original was stationed at Ogden, Utah, and called upon to mark the completion of the first transcontinental railroad in May 1869. It served the route until 1903.

◁ **CP No. 60 *Jupiter*, 1868**

Wheel arrangement 4-4-0

Cylinders 2

Boiler pressure 110 psi (8 kg/sq cm)

Drive wheel diameter 60 in (1,524 mm) later changed to 61 in (1,600 mm)

Top speed approx. 45 mph (72 km/h)

Jupiter was built in New York, shipped in kit-form to San Francisco via Cape Horn, then transported by barge to Sacramento, where it was reassembled. The locomotive represented the Central Pacific Railroad at the "golden spike" ceremony on completion of the transcontinental railroad. This replica was built in 1979.

Thatcher Perkins

Designed by the Master of Machinery at the Baltimore & Ohio (B&O) Railroad, Thatcher Perkins, the Class B No.147 was built in 1863. It entered service the same year, and was used to transport Union troops during the Civil War. Subsequently, No.147 hauled freight and passenger trains in West Virginia until its retirement in 1892. It was given the name *Thatcher Perkins* for the B&O's centennial celebrations in 1927.

WITH EXTRA GRIP from its 4-6-0 wheel arrangement, No.147 was a natural progression from the 4-4-0 locomotive workhorses first used by US railroads. Equipped with Stephenson link-motion valve gear and a large, spark-arresting smokestack and oil lamp, this 40-ton locomotive was designed to pull first-class passenger trains on the company's steeply graded line from Cumberland to Grafton in what is now West Virginia. It replaced a similar locomotive destroyed in the Civil War in 1861, and began service hauling Union troops and munitions across the Allegheny Mountains during the war.

The locomotive's heavy build kept it in service for 29 years, after which it was retired and preserved by B&O for exhibitions and other public-relations purposes. Since 1953, *Thatcher Perkins* has been on display in the Mount Clare Roundhouse at the B&O Railroad Museum in Baltimore, MD. In 2003, the building's roof collapsed during a blizzard and the locomotive was seriously damaged. It has since been restored, and is now back on display in the museum.

FRONT VIEW   **REAR VIEW**

B. & O. R. R.

Leading the way
Opening in 1827, the Baltimore & Ohio Railroad was the first railroad in the US to operate scheduled freight and passenger services for the public.

SPECIFICATIONS			
Class	B	In-service period	1863-92 (Thatcher Perkins)
Wheel arrangement	4-6-0	Cylinders	2
Origin	United States	Boiler pressure	175 psi (12.3 kg/sq cm)
Designer/builder	Thatcher Perkins/B&O	Drive wheel diameter	60 in (1,524 mm) as built
Number produced	11 Class B	Top speed	approx. 50 mph (80 km/h)

Tender is carried on two four-wheel trucks

Handbrake controls brakes on rear truck of tender only

Engineer's cab made of seasoned hardwood

Firebox dome contains safety valve and throttle

Warning bell controlled from cab by a cord

Chimney equipped with spark arrestor

Pilot deflects objects away from track

Safety first
When first introduced, No.147 burned enormous quantities of wood, which it carried in the eight-wheel tender. The locomotive's spark-arresting chimney had a double layer of mesh that stopped wood embers from floating away and setting fire to trackside buildings and vegetation.

EXTERIOR

B&O painted No. 147 in a bold color scheme. The large headlight, pilot (cowcatcher), sand dome, driving cab, and tender body were finished in Indian red with gold lettering and lining. The locomotive and tender wheels as well as the top of the smokestack were painted vermilion, while the cylinders, smoke box, chimney, wheel splashers, underparts, and boiler casing were black. Finally, the boiler bands, flag holders, bell. and oil cups were made of brass. Unlike in Europe, it was common practice in North America to add pilots, or cowcatchers, to the front of steam locomotives to deflect obstacles from the track.

1. Engine number plate **2.** Pilot (cowcatcher) **3.** Headlight **4.** Oil cup for lubricating steam chest **5.** Cylinder housing piston **6.** Linkage for valve gear **7.** Brass bell with decorative mounting yoke **8.** Sand dome **9.** Brass whistle **10.** Oil cups for lubricating side rods **11.** Driving wheels and side rods **12.** Tender truck **13.** Crosshead **14.** Cab windows **15.** Steps up to tender **16.** Link-and-pin coupler at rear of tender

CAB AND TENDER

The spacious engineer's cab was built of wood
and protected the engineer and fireman from
the elements; the cab also featured arched
windows allowing a good view of the track
ahead. Cords to operate the whistle and bell
hung from the roof, while seats were arranged
at each side of the firebox door and offered
the crew a touch of comfort.

Early American steam locomotives used
vast quantities of wood carried in a large
tender at the rear. No. 147's tender, which also
contained a water tank taking up the two sides
and rear, was carried on two four-wheel trucks.

17. Locomotive cab (rear view) **18.** Water level gauge
(sight glass) **19.** Boiler pressure gauge **20.** Water
tricocks **21.** Firebox doors **22.** Johnson bar (reverser
bar) **23.** Fireman's seat **24.** Handbrake wheel
25. Tender coal bunker

Building Great Railroads
Union Pacific

Completed in 1872, the first transcontinental railroad across North America linked Chicago with California. Today the route is owned by Union Pacific, North America's largest Class 1 freight railroad.

THE UNION PACIFIC RAILROAD (UP) started life in 1862 when President Lincoln signed the Pacific Railraod Act authorizing the building of the first transcontinental railroad across North America. Following the wagon-train trails made by pioneer emigrants heading west, the UP was to build westward from Omaha, Nebraska, on the west bank of the Missouri River. The Central Pacific Railroad (CP) was to build eastward from Sacramento, California.

The CP began laying track from Sacramento in 1863. All of the equipment for this section had first to be brought on a long and often dangerous voyage around Cape Horn from the East Coast, a journey that could sometimes take several months.

Union Pacific freight train
The Union Pacific owns nearly 95,000 freight cars and operates double-stack intermodal freight over its 31,800 miles (51,177 km) of track between the East and West coasts.

Travel poster
A woman overlooks a lush California valley in this promotional poster for Union Pacific from around 1915.

Led by the railroad's first General Manager, Thomas Durant, and with a workforce of Irish navvies, the UP began building westward along the Platte River valley from Omaha in 1865. Railroad equipment was first delivered for the UP by riverboats. However, the opening of the Chicago, Iowa & Nebraska Railroad (later, the Chicago & North Western Railroad) linking Chicago to Council Bluffs on the

CALIFORNIA CALLS YOU

UNION PACIFIC

UNITED STATES
OF AMERICA

Construction train, 1868 [1]
Construction teams simultaneously worked on east, central, and western sections of the Overland route, as it was known.

Promontory Summit The Golden Spike was driven by officials of the Union Pacific and the Central Pacific at the line's inauguration in 1869.

[3] **Dale Creek Bridge**
The longest trestle on the Union Pacific, completed in 1868, Dale Creek Bridge was 150 ft (46 m) high and was so slender that it swayed in strong winds. It was later bypassed.

Donner Pass The completion of the 1,659-ft (506-m) Tunnel No. 6 in 1868 allowed the Central Pacific to pass through the Sierra Nevada Mountains.

WYOMING

Omaha-Promontory Summit Union Pacific began building this section in 1865.

Promontory Point

Rawlins

NEBRASKA

Ogden

Sacramento

NEVADA

Laramie   Cheyenne

Grand Island

San Francisco

Sacramento-Promontory This section was built by Central Pacific.

Supply train in Utah [2]
Wagons drawn by oxen were used to bring supplies for the construction of the railroad near Echo Canyon, northeast Utah.

Sherman Summit This was the highest point of the track at 8,015 ft (2,443 m). It was later bypassed by a new, lower route.

Sacramento-Oakland (San Francisco) This section was built by Western Pacific.

CALIFORNIA

UTAH

COLORADO

Bailey Railroad Yard [6]
Union Pacific's Bailey Railroad Yard, North Platte, Nebraska, is the world's largest marshaling yard.

Union Pacific diesels
The Union Pacific owns just over 8,000 diesel-electric locomotives, one of which is seen here at the head of a freight train on the Overland route through the California desert.

east bank of the Missouri River opposite Omaha, allowed materials to be delivered by train, and by 1868 this section had reached Sherman Summit.

Meanwhile, in the west, the CP employed 12,000 Chinese laborers to construct 15 tunnels to reach Donner Pass by 1868.

The two lines met on May 9, 1869 at Promontory, where ceremonial golden spikes were driven into the final wooden railroad tie. However, the transcontinental railroad was only finally completed in 1872 when the Union Pacific opened its bridge across the Missouri River, linking Omaha and Council Bluffs.

Passenger trains were discontinued in 1971 when the newly formed Amtrak took over responsibility for these services. Today, apart from a daily passenger service aboard the luxurious *California Zephyr*, the route carries only freight.

KEY FACTS

DATES
1863 First Central Pacific rails laid in Sacramento
1865 First Union Pacific rails laid in Omaha
1869 Golden spike ceremony at Promontory
1872 Missouri River Bridge completes the line
1883 First passenger service on *Overland Flyer*

TRAINS
First steam locomotive 4-4-0 *Major General Sherman* built in 1864, first saw service in 1865
Largest steam locomotive 4000-Class 4-8-8-4 articulated locomotives or "Big Boys," 1941–44
Diesel-electric locomotives Union Pacific currently operates 8,000, including General Electric 4,400 hp CC44AC/CTE; EMD 4,000 hp SD70M

JOURNEY
Chicago to San Francisco 2,300 miles (3,700 km)
1893 *Overland Flyer* takes 86 hours 30 minutes including a ferry from Oakland to San Francisco
1906 The *Overland Limited* takes 56 hours
Current travel time 51 hours

RAILROAD
Gauge 4 ft 8 ½ in (1,435 mm)
Tunnels Union Pacific: 4; Central Pacific: 15; longest is Summit Tunnel 1,750 ft (533 m)
Longest bridge Dale Creek Bridge 600 ft (183 m)
Highest point Sherman Summit 8,015 ft (2,443 m); now bypassed

A JOINT EFFORT
Although named the Union Pacific Railroad, the route was originally built by four companies: the Chicago, Iowa & Nebraska Railroad; the Union Pacific; the Central Pacific; and the Western Pacific.

KEY
● Start/Finish
● Main stations
Union Pacific
Central Pacific
Chicago, Iowa & Nebraska
Western Pacific

4 The Overland train
The *Overland Flyer*, later the *Overland Limited*, ran part of the Union Pacific route from 1887. Passengers could enjoy the scenery from the observation car at the train's rear.

Chicago–Council Bluffs (Omaha)
This section of the line, built by Chicago, Iowa & Nebraska Railroad, predated the other sections of the Union Pacific Railroad.

5 *Forty-Niner* to San Francisco
The *Forty-Niner*, named for the miners of the California Gold Rush in 1849, was a heavyweight steam streamliner that departed five times a month from Chicago in the 1940s.

IOWA
Chicago
Omaha
ILLINOIS

Council Bluffs A ferry transferred passengers across the Missouri River to Omaha before the bridge was built.

0 150 300 miles
0 150 200 450 km

British Advances

This period of British railroad history features both successes and failures. The famous Grand Junction Railway's Crewe Works opened on a greenfield site in 1840 and was soon turning out graceful single-wheeler express locomotives. While in Liverpool, Edward Bury pursued his bar-frame design, which became popular in North America. On the downside, Brunel's atmospheric railway in Devon was an unmitigated disaster, and the failure of John Fowler's underground steam locomotive caused the designer much embarrassment.

△ FR No.3 "Old Coppernob," 1846

Wheel arrangement	0-4-0
Cylinders	2 (inside)
Boiler pressure	100 psi (7 kg/sq cm)
Drive wheel diameter	57 in (1,448 mm)
Top speed	approx. 30 mph (48 km/h)

Nicknamed "Old Coppernob" because of the copper jacket around its firebox, this locomotive was designed by Edward Bury, and built at Bury, Curtis & Kennedy of Liverpool for the Furness Railway in northwest England. It is normally at the National Railway Museum, York, and is the only survivor of the bar-frame design in the UK.

△ Fireless locomotive "Fowler's Ghost," 1861

Wheel arrangement	2-4-0
Cylinders	2 (inside)
Boiler pressure	160 psi (11.25 kg/sq cm)
Drive wheel diameter	72 in (1,830 mm)
Top speed	approx. 20 mph (32 km/h)

This experimental locomotive, designed by John Fowler and built by Robert Stephenson & Co., was intended for use on London's broad-gauge Metropolitan underground railroad. The engine was equipped with condensing apparatus to prevent steam and smoke emissions. It was a complete failure.

▷ GJR *Columbine*, 1845

Wheel arrangement	2-2-2
Cylinders	2
Boiler pressure	120 psi (8.43 kg/sq cm)
Drive wheel diameter	72 in (1,830 mm)
Top speed	approx. 40 mph (64 km/h)

The locomotive *Columbine*, designed by Alexander Allen, was the first to be built at the Grand Junction Railway's Crewe Works. It was subsequently used to pull the London & North Western Railway's Engineering Department inspection saloon. It hauled passenger trains until 1877 and was withdrawn in 1902. It is now a static exhibit at London's Science Museum.

◁ **FR *Prince*, 1863**

Wheel arrangement 0-4-0ST

Cylinders 2

Boiler pressure 160 psi (11.25 kg/sq cm)

Drive wheel diameter 24 in (610 mm)

Top speed approx. 20 mph (32 km/h)

Businessman and engineer George England designed and built *Prince*. It was one of the first three steam locomotives delivered to the slate-carrying 1-ft 11½-in- (0.60-m-) gauge Ffestiniog Railway in North Wales in 1863. It was returned to service in 2013 to mark the 150th anniversary of steam on the railroad, and is the line's oldest working engine.

▷ **LSWR Class 0298, 1863**

Wheel arrangement 2-4-0WT

Cylinders 2

Boiler pressure 160 psi (11.25 kg/sq cm)

Drive wheel diameter 67 in (1,702 mm)

Top speed approx. 40 mph (64 km/h)

The Class 0298 was designed by Joseph Beattie for the London & South Western Railway to provide suburban passenger services in southwest London. A total of 85 of these well-tank locomotives were built, the majority by Beyer Peacock & Co.

Brunel's Atmospheric Railway

British engineer Isambard Kingdom Brunel built the broad-gauge South Devon Railway between Exeter and Totnes as an "atmospheric railway." Dispensing with locomotives, trains were pushed along by a long piston enclosed in a cast-iron tube in the middle of the track. The vacuum to move the piston was created at stationary pumping houses (such as the one above). The railroad opened in 1847, but failed within a year. In 1848 it was converted to operate with conventional haulage, because the grease that was applied to the leather flap that sealed the pipe melted during hot weather, or was eaten by rats.

Atmospheric railroad track This section of Brunel's broad-gauge track with its cast-iron vacuum pipe is on display at Didcot Railway Centre.

◁ **LNWR *Pet*, 1865**

Wheel arrangement 0-4-0ST

Cylinders 2 (inside)

Boiler pressure 120 psi (8.43 kg/sq cm)

Drive wheel diameter 15 in (380 mm)

Top speed approx. 5 mph (8 km/h)

John Ramsbottom, the locomotive superintendent of the London & North Western Railway, designed this engine. *Pet* is a small cabless steam locomotive that worked on the 1-ft-6-in- (0.45-m-) narrow-gauge Crewe Works Railway until 1929. It is now a static exhibit at the National Railway Museum, York.

Euro Progress

The 1840s saw rapid railroad building across Europe, with many locomotive designs still heavily influenced by the expertise of British engineers; many had set up workshops in France and Austria. By the 1850s, Thomas Crampton's unusual long-boilered, "single-wheeler" engines were hauling trains between Paris and Strasbourg at speeds exceeding 70 mph (113 km/h). The design and craftsmanship of locomotives built by fledgling European builders such as Strauss of Munich stood the test of time, with many remaining in service well into the 20th century.

▷ **Oldenburgische Class G1 No.1 *Landwührden*, 1867**

Wheel arrangement	0-4-0
Cylinders	2
Boiler pressure	142 psi (9.98 kg/sq cm)
Drive wheel diameter	59 in (1,500 mm)
Top speed	37 mph (60 km/h)

The first locomotive to be built by Georg Krauss of Munich, No.1 *Landwührden* won a gold medal for excellence and design of workmanship at the World Exhibition in Paris in 1867. After first working on the Grand Duchy of Oldenburg State Railways' branch lines, this lightweight engine was retired in 1900 and is now on display at the Deutsches Museum in Munich.

▽ **SNB *Limatt*, 1847**

Wheel arrangement	4-2-0
Cylinders	2
Boiler pressure	85 psi (6 kg/sq cm)
Drive wheel diameter	59 in (1,500 mm)
Top speed	approx. 35 mph (56 km/h)

Built by Emil Kessler of Karlsruhe, Germany, *Limatt* was the first steam locomotive on the Swiss Northern Railway (Schweizerische Nordbahn, or SNB), Switzerland's first railroad. The engine is named after the Limmat River, which the railroad followed for much of its route. It is on display at the Swiss Museum of Transport in Lucerne.

▷ **CF de l'Est Crampton, 1852**

Wheel arrangement	4-2-0
Cylinders	2
Boiler pressure	120 psi (8.43 kg/sq cm)
Drive wheel diameter	84 in (2,134 mm)
Top speed	79 mph (127 km/h)

These fast locomotives, designed by British engineer Thomas Crampton, featured a large drive wheel at the rear and a low-mounted boiler. Built by Jean-François Cail, No.80 *Le Continent* hauled express trains between Paris and Strasbourg, retiring only in 1914 after covering 1.5 million miles (2.4 million km).

◁ **Südbahn Class 23 GKB 671, 1860**

Wheel arrangement	0-6-0
Cylinders	2
Boiler pressure	98 psi (6.89 kg/sq cm)
Drive wheel diameter	49 in (1,245 mm)
Top speed	28 mph (45 km/h)

This engine was built by the Lokomotivfabrik der StEG of Vienna to haul freight trains on the Graz-Köflacher Railway in southern Austria. Still used to haul excursion trains, GKB 671 is the oldest steam locomotive in continuous use in the world.

△ **CF de l'Ouest Buddicom Type 111 No. 33 *Saint-Pierre*, 1844**

Wheel arrangement	2-2-2
Cylinders	2
Boiler pressure	80 psi (5.62 kg/sq cm)
Drive wheel diameter	75 in (1,905 mm)
Top speed	37 mph (60 km/h)

Built in Rouen, France, by British engineer William Buddicom for the new Paris to Rouen railroad, No. 33 *Saint-Pierre* had a long and successful career, retiring only in 1912. It is the oldest original steam locomotive still preserved on the European mainland, and is on display at the Cité du Train Museum in Mulhouse.

◁ **BG Type 1B N2T *Muldenthal*, 1861**

Wheel arrangement	2-4-0
Cylinders	2
Boiler pressure	110 psi (8 kg/sq cm)
Drive wheel diameter	48 in (1,220 mm)
Top speed	30 mph (48 km/h)

Sächsische Maschhinenfabrik of Chemnitz built the Type 1B N2T *Muldenthal* to haul coal trains on the newly opened Bockwaer Railway in Saxony. When retired in 1952 it was the oldest operational locomotive in Germany. It is now on display at the Dresden Transport Museum.

TALKING POINT

Class Travel

From the very early days, rail passengers were sorted according to their ability to pay and their position in society. While first-class passengers got sumptuous seating and plenty of space, second class was often overcrowded and the seats were generally wooden. Those in third class traveled in uncovered wagons open to the elements and to the smoke, cinders, and ash from the steam engine at the front.

FIRST CLASS

SECOND CLASS

THIRD CLASS

A Day at the Races, **1846** This cartoon from the *London Illustrated News* shows the social distinctions of class travel on British railroads.

Isambard Kingdom Brunel 1806-59

Audacious and controversial, Isambard Kingdom Brunel became Great Britain's most innovative and successful engineer of the Victorian era. As a young man, he worked on designs for bridges and commercial docks with his father Marc, a French emigré inventor and engineer. Brunel's career took off in 1826 when he was appointed resident engineer for the Thames Tunnel program between Wapping and Rotherhithe in London. Besides designing several of Britain's most famous railroads, bridges, viaducts, and tunnels, Brunel was also involved in several dock plans and three designs for transatlantic ships, which together transformed the face of Victorian England.

GREAT WESTERN RAILWAY

Brunel's most enduring contribution to railroad development in Great Britain was the Great Western Railway (GWR), linking Bristol with London. Despite having no previous experience of rail engineering, he was selected for what was the most technically challenging civil engineering project of its time.

Building began simultaneously from both the London and Bristol termini in 1835, and the line opened in 1841. The 118-mile (190-km) route became famous for its smooth ride, and earned the GWR the nickname "Brunel's billiard table." Brunel's desire to establish the GWR as the fastest and most comfortable line saw him adopt a broad rail gauge of 7 ft ¼ in (2.14 m) instead of George Stephenson's standard gauge of 4 ft 8½ in (1.44 m), which had been used on rail lines in the Midlands and the North. It led to the "Battle of the Gauges," which lasted 50 years, until the GWR finally embraced the standard gauge in 1892. The line remains a key route on Britain's rail network.

Brunel's final engineering project was the Royal Albert Bridge, famous for its gigantic tubular arches. By the time it was complete in 1859, Brunel was too ill to attend the opening, but managed to view his imposing masterpiece by lying on a platform truck that was hauled slowly across the bridge.

Box Tunnel
Built in 1841, Box Tunnel in Wiltshire linked the final section of the GWR between Chippenham and Bath. The construction of the 2-mile- (3.2-km-) long tunnel claimed the lives of more than 100 laborers.

Reaching Cornwall
Brunel built the Royal Albert Bridge (1859) to carry trains across the Tamar River into Cornwall, extending the network westward. Here, a prefabricated bridge span is being prepared to be raised into position.

The GWR's Broad Gauge

While other British railroads were being built to the standard gauge of 4 ft 8½ in (1.435 m),
engineer Isambard Kingdom Brunel used the broad gauge of 7 ft ¼ in (2.14 m) when
building the Great Western Railway, which opened its line from London Paddington
to Bristol in 1841. Brunel had successfully argued that his design offered higher speeds,
smoother running, more stability, and increased comfort for passengers compared to
standard-gauge railroads. In many ways he was right, but the spread of the standard
gauge not only in Britain but also in many other parts of the world, including
North America, made Brunel's broad gauge an anachronism.
The GWR's last broad-gauge train ran on May 21, 1892.

◁ GWR Firefly Class
Fire Fly, 1840

Wheel arrangement	4-2-2
Cylinders	2 (inside)
Boiler pressure	100 psi (7 kg/sq cm)
Drive wheel diameter	84 in (2,134 mm)
Top speed	approx. 58 mph (93 km/h)

Designed by Daniel Gooch, Firefly was one of
61 express passenger locomotives built for the
Great Western Railway by various builders
between 1840 and 1842. The class was known
for its speed, with the original *Fire Fly* traveling
from Twyford to Paddington in only 37 minutes.
Built in 2005, this working replica is the 63rd
Fire Fly. It operates at Didcot Railway Centre.

◁ GWR Iron Duke Class
Iron Duke, 1846

Wheel arrangement	4-2-2
Cylinders	2 (inside)
Boiler pressure	100 psi (7 kg/sq cm)
Drive wheel diameter	96 in (2,440 mm)
Top speed	approx. 77 mph (124 km/h)

Twenty-nine Iron Duke Class express
passenger locomotives, designed by
Daniel Gooch, were built at the Swindon
Works of the Great Western Railway
and Rothwell & Co. of Bolton-le-Moors
between 1846 and 1855. The working
replica *Iron Duke*, seen here, was
built in 1985 and is on display at
Didcot Railway Centre.

IRON DUKE

◁ **GWR Iron Duke Class** *Sultan*, **1857**

Wheel arrangement	4-2-2
Cylinders	2 (inside)
Boiler pressure	100 psi (7 kg/sq cm)
Drive wheel diameter	96 in (2,440 mm)
Top speed	approx. 77 mph (124 km/h)

One of the Great Western Railway's Iron Duke Class express locomotives, *Sultan* was originally built in 1847, but was involved in an accident at Shrivenham a year later when it ran into a freight train. The prototype of this class, *Great Western*, was originally equipped with one pair of carrying wheels at the front as a 2-2-2. As with other members of the class, *Sultan*'s drive wheels had no flanges to allow movement on curves.

▷ **GWR Iron Duke Class**
Lord of the Isles, **1851**

Wheel arrangement	4-2-2
Cylinders	2 (inside)
Boiler pressure	140 psi (10 kg/sq cm)
Drive wheel diameter	96 in (2,440 mm)
Top speed	approx. 77 mph (124 km/h)

Another express passenger locomotive designed by Daniel Gooch for the Great Western Railway, *Lord of the Isles* was an improved version of the Iron Duke Class with higher boiler pressure, sanding gear, and a better engineer's cab. When new, it was exhibited at the Great Exhibition of 1851, and then in Chicago in 1893. It was withdrawn in 1884.

◁ **GWR Rover Class, 1870/1871**

Wheel arrangement	4-2-2
Cylinders	2 (inside)
Boiler pressure	145 psi (10.19 kg/sq cm)
Drive wheel diameter	96 in (2,440 mm)
Top speed	approx. 77 mph (124 km/h)

Built between 1871 and 1888, the Great Western Railway's Rover Class of express locomotives was similar to the Iron Duke Class, but with a small increase in boiler pressure and more protective engineer's cabs. They used names previously carried by Iron Dukes and stayed in service until the end of the broad gauge in 1892.

TECHNOLOGY

Battle of the Gauges

There were major problems for passengers who were forced to change trains at stations where the Great Western Railway's broad gauge met standard-gauge tracks. In 1846 the British Government passed the Railway Regulation (Gauge) Act, which mandated the 4-ft 8½-in (1.435-m) gauge for UK and 5 ft 3 in (1.6 m) for Ireland. Brunel was overruled, and by 1892 all the GWR's lines were converted to standard gauge.

***Break of Gauge at Gloucester*, 1846** This political cartoon depicts the confusion caused at Gloucester station where passengers with luggage had to change trains from the broad-gauge Great Western Railway to the standard-gauge Midland Railway and vice versa.

△ **GWR Broad Gauge Coach, 1840**

Type	6-wheel, Second Class
Capacity	48 passengers
Construction	iron chassis, wooden coach body
Railroad	Great Western Railway

This replica of a Great Western Railway, broad-gauge, second-class car was built by London's Science Museum to run with their replica *Iron Duke* locomotive, to celebrate the anniversary of the railroad in 1985. It now operates with *Fire Fly* at Didcot Railway Centre.

Mass Movers

As railroads expanded, so did their roles and thus the need for engines designed for specific purposes. British express passenger engines had large drive wheels, which increased the distance traveled in each rotation. For goods trains, haulage power was transmitted through six, eight, or ten smaller wheels, which provided the track adhesion needed for heavy loads. Suburban passenger services stayed on schedule by using tank engines that could run equally well smokebox- or bunker-first. For branch line and switching engines, size and weight were key factors, so the short wheelbase 0-4-0 and the 2-4-0 and 0-6-0 types were preferred.

◁ S&DR No. 25 *Derwent*, 1845

Wheel arrangement	0-6-0
Cylinders	2
Boiler pressure	75 psi (3.29 kg/sq cm)
Drive wheel diameter	48 in (1,220 mm)
Top speed	approx. 10–15 mph (16–24 km/h)

From the middle of the 19th century, the six-wheel goods engine became the principal British locomotive. One of the earliest, Timothy Hackworth's *Derwent* of 1845, served the Stockton & Darlington Railway, in northeast England, until 1869.

▷ Met Class A No. 23, 1864

Wheel arrangement	4-4-0T
Cylinders	2
Boiler pressure	120.13 psi (8.46 kg/sq cm); later 150 psi (10.53 kg/sq cm)
Drive wheel diameter	60½ in (1,537 mm)
Top speed	approx. 45 mph (72 km/h)

Tank locomotives built by Beyer Peacock of Manchester were the mainstay of London's Metropolitan Railway from the 1860s until the advent of electrification. To cut pollution, exhaust steam was returned to the water tanks, where it was condensed for reuse.

Freight and Passenger Cars

Unsurprisingly, the designs of the earliest railroad vehicles were based on proven ideas. Cars adopted the design of the road coach; wagons were no more than enlarged versions of the iron and wooden, four-wheel tubs that had been used in mines for centuries. However, increasing loads—both passenger and goods—faster speeds, and the call for greater comfort and facilities brought about rapid advances.

△ SH Chaldron Wagon, 1845–55

Type	Bucket-type coal wagon
Weight	3.3 tons (3 tonnes)
Construction	Iron platework and chassis
Railroad	South Hetton Colliery

The design of the chaldron—a medieval measure used for weighing coal—was adopted for the earliest type of wagon. This one was used on George Stephenson's railroad at the South Hetton Colliery, County Durham, which opened in 1822.

◁ **LNWR "Large Bloomers," 1851**

Wheel arrangement	2-2-2
Cylinders	2
Boiler pressure	100 psi (7 kg/sq cm); later 150 psi (10.53 kg/sq cm)
Drive wheel diameter	84 in (2,134 mm)
Top speed	approx. 50-60 mph (80-96 km/h)

Designed by James McConnell, 74 of these single-wheeler passenger engines were built for the London & North Western Railway up to 1862. They mainly worked between London and Birmingham. The nickname "Large Bloomers" is attributed to American reformer Amelia Bloomer, who scandalized Victorian society by wearing trousers.

▷ **S&PR No. 5 *Shannon*, 1857**

Wheel arrangement	0-4-0WT
Cylinders	2
Boiler pressure	120 psi (8.43 kg/sq cm)
Drive wheel diameter	35 in (889 mm)
Top speed	approx. 10-12 mph (16-19 km/h)

London's George England & Co. built this well tank for the Sandy & Potton Railway in Bedfordshire. In 1862 *Shannon* was sold to the London & North Western Railway, spending 16 years as a works switcher before ending its career on the Wantage Tramway in Oxfordshire.

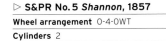

◁ **L&BR Queen Adelaide's Saloon No. 2, 1842**

Type	Passenger car with fold-down beds
Capacity	10 passengers
Construction	Wooden body, iron chassis
Railroad	London & Birmingham Railway

This "stagecoach on wheels" transported Adelaide, Queen Consort to Britain's William IV. While the chassis was entrusted to the London & Birmingham Railway's Euston Works, the body was the work of a London coach builder. This is the oldest preserved car in Europe.

△ **NBR Dandy Car No. 1, 1863**

Type	Horse-drawn rail car
Capacity	30 passengers (12 first and second class, 18 third class)
Construction	Wooden body and frame
Railroad	North British Railway

Between 1863 and 1914, passengers on the Port Carlisle Railway in northwest England traveled in this horse-drawn Dandy Car, the horse trotting between the rails. First- and second-class passengers sat inside, while third class sat on benches at either end.

Building the Tube

Congestion on London's roads was a problem even during the mid-19th century. Charles Pearson, a lawyer, decided to tackle the issue and was instrumental in raising the £1.3 million (about US$6 million) required to build the world's first underground rail line, the 3.75-mile- (6-km-) long Metropolitan. The line would link the City of London in the east and the Great Western Railway station at Paddington in the west, with intermediate stations at King's Cross and Euston.

Construction of the line alternated between open cuttings and tunnels, the latter mostly formed using a "cut-and-cover" method. This involved removing the street surface, cutting a trench, installing the retaining walls, track, and tunnel roof, and finally restoring the street surface.

INSTANT POPULARITY

Londoners immediately took to the underground line. On the opening day, January 10, 1863, 38,000 people rode in wooden-bodied, gas-lit carriages pulled by steam locomotives. Although the exhaust made conditions in stations unpleasant, it did not deter 9.5 million people from using the service in the first year.

The Metropolitan expanded 50 miles (80 km) to the north, but in London the future lay with deep-level lines, electric power, and narrower tunnels—what would become known as the Tube. The first deep-level Tube line opened with electric trains in 1890.

The tunneling shield bore through soft, unstable soil such as clay and acted as a barrier and support while earth and rocks were removed.

Nations and Colonies

The success of the early British railroads and steam engines attracted interest from across Europe and North America. As a result, newly industrialized countries such as the United States, France, and Germany began to lay the foundations for their own national systems and became less and less dependent on British expertise.

However, Britain had a wider sphere of influence: its empire. The first railroad outside Europe and North America was built in the British colony of Jamaica. There were both economic and political reasons for the British to build railroads in Australia, Canada, South Africa, and elsewhere. The vastness of India was controlled through its railroad system, while the efficiency, and therefore profitability, of its mining, logging, and agriculture was completely transformed by the new transportation.

DIE ERSTE BORSIG-LOKOMOTIVE
AUS DEM JAHRE 1841

△ **Borsig No. 1, 1840**

Wheel arrangement	4-2-2
Cylinders	2
Boiler pressure	80 psi (5.62 kg/sq cm)
Drive wheel diameter	54 in (1,372 mm)
Top speed	approx. 40 mph (64 km/h)

August Borsig opened a factory in Berlin in 1837 and three years later delivered his first locomotive to the Berlin–Potsdam Railway. In 1840, No. 1 outpaced a British-built competitor, ending Germany's reliance on imports and helping make Borsig one of the world's leading engine builders.

△ **I-class No. 1, 1855**

Wheel arrangement	0-4-2
Cylinders	2
Boiler pressure	120 psi (8.43 kg/sq cm)
Drive wheel diameter	66 in (1,676 mm)
Top speed	approx. 20 mph (32 km/h)

One of four I-Class locomotives built by Robert Stephenson & Co of Newcastle-upon-Tyne, England, No. 1 was delivered to the Sydney Railway Co. in January 1855. Train services were inaugurated in Australia that May. No. 1 was retired in 1877, having run 156,542 miles (250,467 km).

▷ **EIR No. 22 *Fairy Queen*, 1855**

Wheel arrangement	2-2-2
Cylinders	2
Boiler pressure	80–100 psi (5.62–7 kg/sq cm)
Drive wheel diameter	72 in (1,830 mm)
Top speed	approx. 25 mph (40 km/h)

The first locomotive to haul passenger trains in India was built by Kitson, Hewitson & Thompson of Leeds, England, in 1855 for the East Indian Railway. An outside-cylinder, 2-2-2 well tank, the *Fairy Queen* is part of India's historic locomotive collection in New Delhi and has a claim to be the world's oldest working locomotive.

△ *La Porteña*, 1857

Wheel arrangement	0-4-0ST
Cylinders	2
Boiler pressure	140–160 psi (9.84–11.25 kg/sq cm)
Drive wheel diameter	about 48 in (1,219 mm)
Top speed	approx 16 mph (26 km/h)

Arriving in Argentina from Britain on Christmas Day, 1856, the outside-cylindered, four-wheel saddletank *La Porteña* hauled the first train over the Buenos Aires Western Railway on August 29, 1857. Built by E. B. Wilson of Leeds, it remained in service until 1899 and is now exhibited at the museum in Luján.

◁ **Hawthorn No. 9 *Blackie*, 1859**

Wheel arrangement	0-4-2
Cylinders	2
Boiler pressure	130 psi (9.14 kg/sq cm)
Drive wheel diameter	54 in (1,372 mm)
Top speed	approx. 30 mph (48 km/h)

Hawthorn & Co. assembled this 0-4-0 at its works in Leith, Scotland, for contractor Edward Pickering, who used it in the construction of the 45-mile (72-km) Cape Town to Wellington Railway. South Africa's first locomotive, it was rebuilt as a 0-4-2 in 1873-74 and is now exhibited at Cape Town's main station.

△ **O&RR Class B No. 26, 1870**

Wheel arrangement	0-6-0
Cylinders	2
Boiler pressure	160-180 psi (11.25-12.65 kg/sq cm)
Drive wheel diameter	52 in (1,320 mm)
Top speed	approx. 40 mph (64 km/h)

This locomotive was built by Sharp, Stewart & Co. of Manchester, England, for the 5-ft-6-in- (1.67-m-) gauge Oudh & Rohilkhand Railway of northern India. No. 26 is typical of British engines exported then.

TECHNOLOGY

Challenging Railroads

With mountain ranges, deserts, and jungles to be overcome, India posed a huge challenge to railroad builders. Nevertheless, the first 25-mile (40-km) stretch between Bombay (now Mumbai) and Thane opened in November 1852, and by 1880 around 9,000 miles (14,484 km) of track had been laid. Twenty years on, the network had extended to 40,000 miles (64,374 km). A committee set up by the Governor General, Lord Dalhousie, led to the creation of the Great Indian Peninsular Railway, the East India Railway, and the famous Darjeeling Himalayan Railway.

Construction site Workers photographed in 1856 on the wooden staging used in the building of the viaduct at the mouth of tunnel No. 8 (out of 28) on the Bhor Ghat Railway.

1870-1894

A WORLD
OF STEAM

A WORLD OF STEAM

When Bombay's Victoria Terminus opened in 1888, it was heralded as one of the world's grandest stations. Owing its styling to elements from both Indian and British history, it had taken 10 years to build. "VT"—now known as Mumbai's Chhatrapati Shivaji Terminus—became symbolic of an era in which nothing seemed beyond human endeavor and ingenuity.

At this time, railroads were spreading across the globe, climbing or boring through mountains and crossing mighty waterways via bridges, or being linked by steamships across vast seas and oceans. In 1881 the narrow-gauge Darjeeling Himalayan Railway opened, running from India's plains high into the foothills of the Himalayas. Meanwhile, the construction of Switzerland's Gotthard Tunnel had pushed a main line through 9 miles (15 km) of mountain rock. In 1885 the Canadian Pacific Railway was completed, creating a second route that spanned an entire continent. In the UK, the Forth Bridge opened in 1890, crossing the Firth of Forth for more than 1½ miles (2.5 km). Then, in 1891, work started on a project that would dwarf almost everything else: Russia's Trans-Siberian Railway would connect Moscow to Vladivostok on the country's distant eastern coast.

There was an insatiable demand for more lines, higher speeds, more luxury, and greater magnificence. The railroads' glamorous and luxurious side was epitomized by the development of the long-distance *Orient Express*, which by 1891 had connected Paris and Constantinople (Istanbul) via some of Europe's most important cities. Yet among all the expansions and improvements to steam travel, there were also early signs of a different future: in 1879, a new electric locomotive, which drew power from the track, was demonstrated in Berlin.

> # "**Lay down your rails**, ye Nations, near and far; Yoke your full trains to **Steam's triumphal car**"
>
> CHARLES MACKAY, SCOTTISH POET

◁ **Japanese onlookers** watch an early US navy demonstration of steam in Yokohama

△ **Rush hour on the "El"**
In the late 19th century the Manhattan Railway Co. operated four elevated lines in New York City.

Key Events

▷ **1870s** The electric "track circuit" is developed; it automatically shows signalers the location of trains.

▷ **1871** New York's Grand Central Station opens—it is later rebuilt as Grand Central Terminal.

▷ **1872** Japan's first railroad opens between Tokyo and Yokohama.

▷ **1879** Werner von Siemens demonstrates an electric locomotive in Berlin; the following year, an electric tramway is trialed in St. Petersburg.

▷ **1881** The narrow-gauge Darjeeling Himalayan Railway is completed, connecting the Darjeeling hill station to India's rail network.

▷ **1883** One of the world's most glamorous trains is launched. From 1891 it is known as the *Orient Express*.

▷ **1885** The Canadian Pacific Railway's transcontinental route is completed.

▷ **1888** Bombay's Victoria Terminus is completed a decade after work started.

△ **Victorian Gothic**
Victoria Terminus, designed by the consulting British architect Frederick William Stevens, bears some resemblance to St. Pancras station in London.

▷ **1888** British railroad companies compete in the London to Edinburgh "Race to the North."

▷ **1890** The Forth Bridge opens, seven years after construction started.

▷ **1891** Work begins on one of the most ambitious engineering projects ever— the Trans-Siberian Railway.

19th-century Racers

The development of sleek express steam engines in the late 19th century led publicity-seeking railroads in both the US and the UK to compete for the fastest travel times on rival intercity routes. In the UK the famous "Races to the North" of 1888 and 1895 saw the railroads of the rival East Coast and West Coast Main Lines between London and Scotland engage in a dangerous, high-speed struggle for supremacy. In the US there was fierce competition between the Pennsylvania Railroad and the New York Central & Hudson River Railroad on their New York to Buffalo routes during the 1890s. This triggered electrifying performances by the latter company's celebrity locomotive No. 999 while hauling the *Empire State Express*.

◁ **GNR Stirling Single Class, 1870**

Wheel arrangement	4-2-2
Cylinders	2
Boiler pressure	170 psi (11.95 kg/sq cm)
Drive wheel diameter	97 in (2,464 mm)
Top speed	85 mph (137 km/h)

Patrick Stirling designed this locomotive for the Great Northern Railway. A total of 53 of these single-wheeler locomotives were built at Doncaster Works between 1870 and 1895. The locomotives hauled express trains on the East Coast Main Line between London King's Cross and York and were involved in the "Races to the North" of 1888 and 1895. No. 1, shown here, is preserved at the National Railway Museum in York, UK.

Races to the North

Headlined in newspapers as the "Race to the North," railroad companies unofficially raced each other on two main lines between London and Edinburgh in 1888. The West Coast Main Line trains were operated by the London & North Western Railway and the Caledonian Railway; and the East Coast Main Line trains by the Great Northern Railway, the North Eastern Railway, and the North British Railway. Following the completion of Scotland's Forth Bridge in 1890, the companies raced between London and Aberdeen. After a derailment at Preston in 1896, the practice was banned and speed limits were enforced.

Record run A Caledonian Railway postcard shows Engine No. 17 and driver John Souter at Aberdeen after their race-winning run on August 23, 1895.

▷ **LNWR Improved Precedent Class, 1887**

Wheel arrangement	2-4-0
Cylinders	2 (inside)
Boiler pressure	150 psi (10.54 kg/sq cm)
Drive wheel diameter	81 in (2,057 mm)
Top speed	approx. 80 mph (128 km/h)

A total of 166 Improved Precedent Class express locomotives, designed by F.W. Webb, were built at the London & North Western Railway's Crewe Works between 1887 and 1901. No. 790 *Hardwicke* set a new speed record between Crewe and Carlisle during the "Race to the North" on August 22, 1895. It is preserved at the National Railway Museum in York, UK.

◁ CR No. 123, 1886

Wheel arrangement	4-2-2
Cylinders	2 (inside)
Boiler pressure	160 psi (11.25 kg/sq cm)
Drive wheel diameter	84 in (2,134 mm)
Top speed	approx. 80 mph (129 km/h)

Built as an exhibition locomotive by Neilson & Co. of Glasgow for the Caledonian Railway in 1886, this unique single-wheeler hauled expresses between Carlisle and Glasgow. Following retirement in 1935 it was preserved and is now on display at the Riverside Museum in Glasgow.

Standard Rail Time

Confusion reigned on the early railroads as clocks at stations were set at local time, causing difficulty for railroad staff and passengers alike. In Great Britain, the Great Western Railway introduced a standardized "London Time" for their station schedules in 1840. This synchronization used Greenwich Mean Time (GMT), set by the Royal Observatory at Greenwich, which later became accepted as the global standard time. In 1883, railroads in the US and Canada split both countries longitudinally into geographic time zones and introduced Railroad Standard Time.

Time regulation Made by American jeweler Webb C. Ball in 1889, this precision regulator clock helped maintain the accuracy of other timepieces on the Baltimore & Ohio Railroad.

▷ NYC&HR No. 999, 1893

Wheel arrangement	4-4-0
Cylinders	2
Boiler pressure	180 psi (12.65 kg/sq cm)
Drive wheel diameter	86½ in (2,197 mm)
Top speed	approx. 86 mph (138 km/h)

Alleged to have traveled at over 100 mph (161 km/h), No. 999 was built in 1893 to haul the New York Central & Hudson River Railroad's flagship train, the *Empire State Express*, between New York and Buffalo. This celebrity locomotive was exhibited at the Chicago World's Fair before being retired in 1952. Nicknamed the "Queen of Speed," No. 999 is on display at the Chicago Museum of Science & Industry.

▽ LB&SCR B1 Class, 1882

Wheel arrangement	0-4-2
Cylinders	2 (inside)
Boiler pressure	150 psi (10.53 kg/sq cm)
Drive wheel diameter	78 in (1,980 mm)
Top speed	approx. 70 mph (113 km/h)

The B1 Class locomotives were designed by William Stroudley for the London, Brighton & South Coast Railway. A total of 36 were built at Brighton Works between 1882 and 1891. Hauling heavy expresses between London and Brighton, they were named after politicians, railroad officials, or places served by the railroad. The last survivor was retired in 1933, and No. 214 *Gladstone* is preserved at the National Railway Museum in York.

London Locals

Growing prosperity and personal mobility enabled people to move away from the city center. Railroads supplied transportation links from the new suburbs to the city, giving birth to the commuter train. While the UK's Great Eastern Railway, among others, provided a rush-hour, steam-hauled service, electric traction—overground and underground— was the future. The first deep-level "tube" line, the City & South London Railway, which opened in 1890 formed the nucleus of London's subway system. Other cities soon followed London's example: Liverpool in northwest England and Budapest and Paris in continental Europe. In the US, the Boston subway opened in 1897 and, by 1904, had been joined by New York's.

△ GWR 633 Class, 1871

Wheel arrangement	0-6-0T
Cylinders	2
Boiler pressure	165 psi (11.6 kg/sq cm)
Drive wheel diameter	54¹/₂ in (1,384 mm)
Top speed	approx. 40 mph (64 km/h)

Designed by George Armstrong and built at Wolverhampton Works, several of the 12-strong 633 Class were equipped with condensing apparatus to take Great Western Railway trains through the tunnels, so gaining the nickname "Tunnel Motors." Much modified, some lasted until 1934.

△ LB&SCR A1 Class, 1872

Wheel arrangement	0-6-0T
Cylinders	2
Boiler pressure	150 psi (10.53 kg/sq cm)
Drive wheel diameter	48 in (1,219 mm)
Top speed	approx. 60 mph (96 km/h)

The London, Brighton & South Coast Railway's suburban network was the domain of William Stroudley's small, six-coupled tanks. Fifty were built between 1872 and 1880, and the bark of their exhaust earned them the nickname of "Terriers." They were named after places they served—in the case of No. 54 *Waddon* (1875), a district near Croydon.

◁ NLR 75 Class, 1879

Wheel arrangement	0-6-0T
Cylinders	2
Boiler pressure	160 psi (11.24 kg/sq cm)
Drive wheel diameter	52 in (1,321 mm)
Top speed	approx. 30 mph (48 km/h)

John C. Park supplied the North London Railway with this switching engine to serve the dock system around Poplar. Thirty were built up to 1905 and, since they rarely left the docks, no coal bunker was installed; fuel was stored on the footplate.

◁ LSWR 415 Class, 1882

Wheel arrangement	4-4-2T
Cylinders	2
Boiler pressure	160 psi (11.25 kg/sq cm)
Drive wheel diameter	67 in (1,702 mm)
Top speed	approx. 45 mph (72 km/h)

Designed by William Adams of the London & South Western Railway, 71 of the 415 Class were built from 1882 to 1885. Put to work on suburban services out of London's Waterloo, three ended their days on the southwest Lyme Regis branch, where their short wheelbase and leading truck were ideal to negotiate the severe curves.

London's Train Cars

Both the Metropolitan and District railways began by using locomotive-hauled cars. However, few offered the upholstered luxury of the Metropolitan's "Jubilee" coach. Most followed the pattern of the District's No. 100, with 10 passengers to each compartment. The distinction between the classes even extended to lighting: first-class travelers enjoyed two gas jets, while second- and third-class passengers made do with one. Conditions improved little with the coming of the City & South London Railway, which became known as the "sardine-tin railway."

◁ C&SLR "Padded Cell," 1890

Type	underground passenger car
Capacity	32 passengers
Construction	wooden body on two 4-wheel trucks
Railroad	City & South London Railway

Tunnel diameter restricted car size on this first "tube" line. Coaches were equipped with high-backed seats, running along the length, and gates at either end to allow passengers on and off. With the only windows being slits above seats, and air entering through roof ventilators, the nickname "padded cell" was appropriate.

△ **C&SLR electric locomotive, 1889**

Wheel arrangement	0-4-0 (Bo)
Power supply	0.5kV DC third rail
Power rating	100 hp (74.60 kW)
Top speed	25 mph (40 km/h)

The first important railroad to use electric traction was the City & South London Railway. When it opened in 1890, the line had six stations and ran from City to Stockwell. Operated by 14 locomotives, one of which—with a train of later steel-bodied cars with full-length windows—is passing Borough Junction in this 1922 photograph.

TALKING POINT

Cemetery Railroad

As London's population doubled in the 19th century, burying the dead became a crisis. The boldest solution came from Sir Richard Broun and Richard Sprye. Their plan involved buying a large piece of land away from the city but with a direct rail link to London. The chosen location was Brookwood, Surrey, 23 miles (37 km) along the LSWR main line out of Waterloo. They envisioned that coffins would be brought to Brookwood either late at night or early in the morning, with mourners traveling by dedicated train services during the day.

Burying the dead The London Necropolis (Greek for "city of the dead") Railway opened in 1854. After the terminus was bombed in 1941, its services never ran again.

▽ **GER S56 Class, 1886**

Wheel arrangement	0-6-0T
Cylinders	2
Boiler pressure	180 psi (12.65 kg/sq cm)
Drive wheel diameter	48 in (1,220 mm)
Top speed	approx. 60 mph (96 km/h)

James Holden designed a small but powerful, six-coupled tank for the Great Eastern Railway's inner-suburban services in 1886. It was equipped with Westinghouse compressed air brakes, ideal where stations were close together. The sole survivor, No. 87 of 1904, is part of Britain's National Collection.

△ **Met C Class, 1891**

Wheel arrangement	0-4-4T
Cylinders	2
Boiler pressure	140 psi (9.84 kg/sq cm)
Drive wheel diameter	66 in (1,676 mm)
Top speed	approx. 60 mph (96 km/h)

The Metropolitan Railway's C Class consisted of just four engines built by Neilson & Co. of Glasgow. After the Met's expansion into Hertfordshire and Buckinghamshire, they hauled trains from the city out to Watford, Amersham, and Aylesbury.

▷ **Met Jubilee Coach No. 353, 1892**

Type	four-compartment, first-class passenger coach
Capacity	32
Construction	original wooden body on later 4-wheel steel chassis
Railway	Metropolitan Railway

This car served the Metropolitan Railway from 1892 until 1907, when it was sold to the Weston, Clevedon & Portishead Light Railway. Restored to mark the railroad's 150th anniversary in 2013, it is now at the London Transport Museum.

△ **DR Coach No. 100, 1884**

Type	four-compartment, third-class passenger car
Capacity	40
Construction	original wooden body on later 4-wheel steel chassis
Railroad	District Railway

The origins of coach No. 100 of the District Railway are uncertain. What is definite is that the body ended up being used as a storage shed. It was rescued, placed on a new chassis, and now runs on the Kent & East Sussex Railway, where a District Railway brown livery was applied.

End of the Great Western Broad Gauge

When the broad-gauge rails of the Great Western Railway (GWR) first came up against those of a narrower-gauge company at Gloucester, England in 1844, passengers were forced to change trains to continue their journey to the north or southwest. The impracticality of mismatched gauges led the UK government to set up a Gauge Committee to examine the issue. While the committee agreed that the GWR offered greater speed and stability (due as much to the excellence of Isambard Kingdom Brunel's railroad and Daniel Gooch's locomotives as to the width of the track), it concluded that the narrower gauge suited its long-term interest. A change to the narrower gauge (now known as standard gauge) became inevitable.

First, a third rail to take standard-gauge trains was laid on the GWR, reaching Paddington in 1861. Gauge conversion began in 1866 and took almost three decades to complete. The final change, along the West of England main line, took place over a weekend in May 1892. It was meticulously planned with 4,200 workmen positioned along the line and prefabricated track sections such as facing points and crossovers ferried to where they were needed. On May 23, 1892, the operation was completed and Brunel's broad gauge was consigned to history.

On May 20, 1892, the last *Cornishman* broad-gauge express left London Paddington for Penzance with Rover Class 4-2-2 *Great Western* at its head.

C&PA Snow Plow

The Coudersport & Port Allegany Railroad (C&PA) Snow Plow is typical of the wooden plows built by the Russell Company of Ridgeway, PA, from the late 19th century onward. Designed to be pushed by one or two steam engines along a single-track line, it was fitted with a broad wedge-shaped blade, which removed snow pack and debris from the track, and a flanger to clear snow from between the rails.

IN WINTER, HEAVY SNOWFALLS and icy conditions regularly closed train tracks in the US and Canada. Faced with a loss of business, the railroad companies started to use wooden wedges attached to the front of locomotives to clear snow from the tracks. By the late 19th century, this makeshift arrangement was superceded by the introduction of separate snow-plow wagons mounted on trucks and pushed by locomotives.

Believed to be the oldest snow plow of its type in existence, the C&PA Snow Plow is a wedge-style model that was built around 1890 under license for the Russell Snow Plow Company by the Ensign Manufacturing Company in Huntingdon, WV. A cupola was installed on the roof directly behind the plow blades to give the crew a view of the tracks ahead. The plow was used on the C&PA until 1945, when it was damaged in an accident. It later became the property of the Wellsville, Addison & Galeton Railroad before being donated to the Railroad Museum of Pennsylvania in 1980, where it has since been restored.

FRONT VIEW

REAR VIEW

Short-line railroad
Opened in 1882, the Coudersport & Port Allegany Railroad (C&PA) was a 32-mile- (50-km-) lumber-carrying short line in Potter and McKean Counties in Pennsylvania. It was abandoned in the early 1970s.

Rear balcony and entrance to cabin

Portholes to let in light

Chimney for crew's stove

Pilot house cupola allows view beyond blades

Steel-reinforced flanger with horizontal and vertical blades

SPECIFICATIONS	
Type	Wedge-type snow plow
Origin	United States
Designer/builder	Russell Co.
Number produced	1
In-service period	c. 1890-1945
Weight	Not known
Construction	Wood and steel
Railroad	Coudersport & Port Allegany Railroad

Blades at work
As the plow was pushed from behind by one or two steam locomotives, the sharp front blade would lift snow off the track before deflecting it to either side with the angled vertical blade.

EXTERIOR

While the main body of the C&PA Snow Plow was constructed of seasoned hardwood, the impressive plow blades were reinforced with steel. The plow was mounted on two four-wheeled trucks, one of which was concealed beneath the front blade housing.

1. Porthole-style windows above top edge of plow **2.** Front coupling bar **3.** Rivets on front edge of plow **4.** Chimney **5.** Round and square windows looking out from the observation level **6.** Journal box access door **7.** Journal box, which contains the journal bearing **8.** Flanger (secondary plow) **9.** Back wheel brake shoe **10.** Truck at rear (arch bar truck) **11.** Deck behind pilot house **12.** Bars across aperture on platform, used as a ladder to access roof **13.** Coupling ring at rear **14.** Angle cock **15.** Coupling at rear

INTERIOR

The pilot house was reinforced at the front end by steel girders to prevent the crew from being crushed during snow-clearing operations. It was equipped with a handbrake, an air pressure gauge, steps for the cupola forward lookout, and a coal-fired safety stove with a flange on top to keep pans from sliding off. The suspended flanger beneath the cabin floor could be raised or lowered either manually or by air pressure, to avoid damaging switches and grade crossings.

16. Pilot house interior **17.** Suspension springs to the flanger **18.** Air reservoir pipes behind steps **19.** Piston to adjust height of flanger **20.** Brake wheel **21.** Base of brake wheel with cog mechanism **22.** Air brake pressure gauge **23.** Coal stove **24.** Pennsylvania Railroad stamp on stove **25.** Decorative door handle

Delivering to America

The South Carolina Railroad began conveying mail as early as 1833, but the first regular mail service in the US did not start until two years later, on the Baltimore and Ohio. In July 1838, Congress approved the use of all rail routes to carry mail, earning the railroad companies significant revenues from the US Post Office Department (USPOD).

In 1862, President Abraham Lincoln approved the building of a 1,928-mile (3,084-km) line between Omaha, Nebraska, and Sacramento, California, to bring people, trade, and a much-needed postal service to the western states. That same year, the government unified mail traffic under the Railway Mail Service, which led to the construction of dedicated Railway Post Office

(RPO) cars. The first permanent RPO service ran between Chicago, Illinois, and Clinton, Iowa. By 1869 the transcontinental railroad was complete, giving postal cars the means to carry mail across the breadth of the US. At their peak, RPO cars covered more than 200,000 miles (320,000 km) on more than 9,000 routes.

The US also ran the world's first postal express, which left New York's Grand Central Station for Chicago on September 16, 1875. It completed the journey in just over 24 hours, and was the forerunner of night mail trains across the world.

The rail junction at night depicted in this Currier & Ives print of 1876 shows some of the express train types that carried mail to major US cities.

Building Great Railroads
Canadian Pacific

Spanning the vast spaces of the prairies and the Rocky Mountains and opened in 1886, the Canadian Pacific Railway was Canada's first transcontinental line, linking Vancouver on the Pacific west coast with Montreal on the St. Lawrence River.

IN 1871 THE GOVERNMENT of the recently formed Dominion of Canada promised the isolated western province of British Columbia that a railroad would be built across the Rocky Mountains within 10 years. The project got off to a slow start, and by 1880 only 300 miles (483 km) of line had been built.

However, in 1881 a group of Canadian businessmen formed the Canadian Pacific Railway (CP) and, with financial assistance and land from the government, took over the unfinished lines and restarted construction work from both the east and the west. Overseen by the new general manager of the railroad, William Cornelius Van Horne, tracklaying in the east began at Bonfield, north of the Great Lakes, and proceeded slowly westward across the remote, sparsely populated

Colonizing Canada
Canadian Pacific offered packages of sea and rail travel to immigrants.

and lake-strewn landscape of the Canadian Shield in Ontario toward Winnipeg. The link connecting the new railroad at Bonfield with the eastern cities of Ottawa and Montreal had already been built by the Canadian Central Railway and the Ontario & Quebec Railway, both of which the CP leased from 1884.

From Winnipeg, construction continued westwards across the vast plains of Saskatchewan to Calgary at the foot of the Rockies. From Calgary, gangs of Chinese laborers built the railroad up into the Rockies through Banff, reaching Kicking Horse Pass in 1884. From here the railroad made a steep descent down the Big Hill before climbing again to cross the Selkirk Range at Rogers Pass.

Climbing the Rocky Mountains
A passenger train with vista dome cars and an observation car follows the Bow River on the CP's scenic route through the Banff National Park in the Canadian Rockies.

Rush to finish [2]
Temporary timber trestle bridges were built in order to rapidly complete the railroad before funds ran out, and were later replaced by more permanent structures.

[3] **The final spike**
At Craigellachie the final spike was driven by Donald A. Smith, completing the line between Montreal and the Pacific.

C A N A D A

● Jasper   ● Edmonton

BRITISH COLUMBIA

ALBERTA

The "Last Best West" This was a phrase used to market the settlement of the prairieland provinces of Saskatchewan and Manitoba, a program made possible by the railroad.

Lake Winnipeg

Snowy conditions [1]
Snow sheds were built in the 1880s to protect the tracks from large snowfalls.

● Banff
Calgary

● Kamloops

● Vancouver

SASKATCHEWAN

MANITOBA

Medicine Hat   Regina

● Winnipeg

[5] **The Big Hill Spiral Tunnels**
In 1906 construction began on the Spiral Tunnels, needed to address the very steep downhill gradient (4.5 percent) on the Big Hill, which saw many a runaway train. It took 1,000 workers two years to complete the tunnels.

New line
Old line
Tunnel

Lower Spiral Tunnel

Kicking Horse River

Upper Spiral Tunnel

U N I T E D   S T A T E S
O F   A M E R I C A

Across the Canadian prairies
A train travels over the prairie near Morse, between Regina and Medicine Hat, in Saskatchewan. Natural gas was discovered in the prairies by workers constructing the line.

To the west of the Rockies, construction continued through the Monashee Mountains before the two lines met at Craigellachie, where a ceremonial last spike was driven in 1885. The entire route was now complete and the first transcontinental train ran between Montreal in the east and Port Moody in the west in 1886. A year later the western terminus was moved to Vancouver. Attracted by a CP package deal, which included passage on a company ship, travel on a company train, and land sold by CP, thousands of immigrants from Europe were soon streaming westward on the new railroad in search of new lives.

The Big Hill, with its treacherously steep gradients, was bypassed in 1909 when the Spiral Tunnels were opened, and the steep gradient up to Rogers Pass was also later bypassed by the opening of the Connaught Tunnel in 1916.

KEY FACTS

DATES
1881 Construction begins at Bonfield
1882 Thunder Bay branch completed
1885 November 3: Last spike on Lake Superior section; November 7: Final spike at Craigellachie, BC
1886 June 28: First transcontinental passenger service leaves Dalhousie Station, Montreal
1909 Spiral Tunnels at Kicking Horse pass open
1978 CP passenger services taken over by Via Rail
1990 *The Canadian* passenger train rerouted over Canadian National Railways route

FIRST PASSENGER TRAIN
Locomotive type American Standard 4-4-0 steam
Cars 2 baggage cars, 1 mail car, 1 second-class coach, 2 immigrant sleepers, 2 first-class coaches, 2 sleeping cars, and a diner

JOURNEY
Montreal to Port Moody (1886) 2,883 miles (4,640 km); 6 days 6 nights
Montreal to Vancouver (1963) 2,888 miles (4.648 km); 69 hours

RAILROAD
Gauge Standard 4 ft 8 ½ in (1,435 mm)
Tunnels Connaught Tunnel 5 miles (8 km); Spiral Tunnels No. 1: 3,153 ft (961 m), and No. 2: 2,844 ft (867 m)
Bridges Stoney Creek Bridge 300 ft (91 m) high
Highest point 5,338 ft (1,627 m) Kicking Horse Pass

KEY
● Start/Finish
● Main stations
IIIIII Main route

PEAKS AND VALLEYS

Construction in the Rocky Mountains was particularly perilous. Workers faced harsh terrain, the threat of forest fires, heavy snowfall, and avalanches as they built the line across deep valleys, up steep gradients, and through rock.

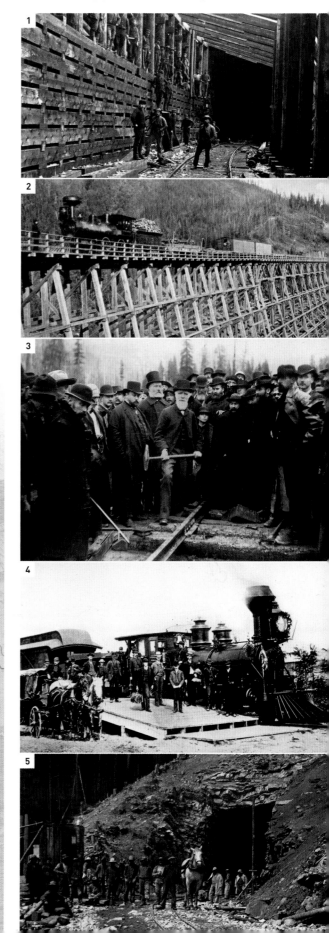

N
0   150   300 miles
0   150   300   450 km

ONTARIO

Muskeg terrain This required sections of the track to be elevated to keep it from sinking during thaws.

Bonfield
This was the site of the first track to be laid.

QUÉBEC

4 First passenger train
The first transcontinental service left Dalhousie Station, Montreal on June 28, 1886.

Thunder Bay

Lake Superior

Montreal

Lake Michigan

Lake Huron

Ottawa

Lake Ontario

Specialty Steam

Initially used for hauling coal, railroads were soon adapted to play similar roles in the fast-growing industrial landscape. Narrow-gauge lines and engines were ideal for quarries, foundries, shipyards, brickworks, and some military sites. Dock railroads required small but powerful engines that could weave their way along docksides, while in chemical plants and munitions factories the danger posed by stray sparks was overcome by developing fireless locomotives. Ingenious engines and track were used to scale mountains. There were few places where the steam locomotive could not serve.

◁ **VRB No. 7, 1873**

Wheel arrangement	0-4-0VBT
Cylinders	2
Boiler pressure	185 psi (13 kg/sq cm)
Drive wheel diameter	25 in (644 mm)
Top speed	approx. 5 mph (8 km/h)

Designed by Niklaus Riggenbach and built by the Swiss Locomotive Co., No. 7 was employed on the Vitznau-Rigi mountain railroad (Vitznau-Rigi Bahn, or VRB) near Lucerne, Switzerland, until 1937. Its vertical boiler kept a safe water level on the steep climb, which was undertaken using a rack-and-pinion system.

▷ **SRR A-4 Class "Camelback," 1877**

Wheel arrangement	0-4-0
Cylinders	2
Boiler pressure	200 psi (14.06 kg/sq cm)
Drive wheel diameter	50 in (1,270 mm)
Top speed	approx. 20 mph (32 km/h)

Engines of Pennsylvania's coal-carrying railroads were fueled by cheap anthracite waste that needed a large firebox for ample combustion, so the engineer's cab could not be sited behind it. Instead, it straddled the firebox, hence the nickname "Camelback." No. 4 worked on the Philadelphia & Reading Railroad and the Strasburg Railroad.

◁ **FR Double Fairlie No. 10**
Merddin Emrys, **1879**

Wheel arrangement	0-4-4-0T
Cylinders	4
Boiler pressure	160 psi (11.25 kg/sq cm)
Drive wheel diameter	32 in (813 mm)
Top speed	approx. 35 mph (56 km/h)

Following a design by British engineer Robert Fairlie, *Merddin Emrys* was the first locomotive built by the Ffestiniog Railway's workshops. A double-ended, articulated tank engine riding on powered trucks, today's No. 10 is much rebuilt.

TECHNOLOGY

Crane Tanks

Used in industrial locations from docks and factories to shipyards and ironworks, crane tanks combined switching with the ability to distribute loads. The Pallion shipyard in Sunderland, in northeast England, employed a fleet of five, while the nearby Shildon Ironworks in County Durham saw the last use of the type in the UK. Crane tanks were chiefly a product of the 19th century, although one—built for the North London Railway—remained in service until 1951.

Southern Railway No. 234S, 1881 This crane tank was used at Ashford Locomotive Works and Folkestone Harbour, both in Kent, and at Lancing Carriage Works in Sussex. It was retired in 1949.

△ LYR *Wren*, 1887

Wheel arrangement	0-4-0ST
Cylinders	2
Boiler pressure	170 psi (11.95 kg/sq cm)
Drive wheel diameter	16¹/₂ in (418 mm)
Top speed	approx. 5 mph (8 km/h)

Wren was one of eight small saddletanks employed on the 7¹/₂-mile (12-km), 1-ft 6-in- (0.46-m-) gauge railroad serving the Lancashire & Yorkshire Railway's works at Horwich, England. The engine was built by Beyer Peacock & Co. of Manchester, and remained in use until 1962.

◁ Hunslet *Lilla*, 1891

Wheel arrangement	0-4-0ST
Cylinders	2
Boiler pressure	120 psi (8.43 kg/sq cm)
Drive wheel diameter	26 in (660 mm)
Top speed	approx. 10–12 mph (16–19 km/h)

Lilla is a survivor from 50 saddletanks built by the Hunslet Engine Co. of Leeds, England, between 1870 and 1932 for Welsh slate quarries. It was retired from Penrhyn Quarry in 1957 and is now preserved on the Ffestiniog Railway in North Wales.

◁ Hunslet *Linda*, 1893

Wheel arrangement	0-4-0STT
Cylinders	2
Boiler pressure	140 psi (9.9 kg/sq cm)
Drive wheel diameter	26 in (660 mm)
Top speed	approx. 12–18 mph (19–29 km/h)

From the same stable as *Lilla* but more powerful, *Linda* was used on the Penrhyn Quarry's "main line," which ran from Bethesda to Port Penrhyn, near Bangor, Wales. Another Ffestiniog veteran, *Linda* has been rebuilt there as a 2-4-0 saddletank tender engine.

▷ Saxon IV K Class, 1892

Wheel arrangement	0-4-4-0T
Cylinders	4 (compound)
Boiler pressure	174 psi/203 psi/217 psi (12.23 kg/sq cm/14.27 kg/sq cm/ 15.25 kg/sq cm) (variations within class)
Drive wheel diameter	30 in (760 mm)
Top speed	approx. 19 mph (30 km/h)

Germany's most numerous narrow-gauge class, 96 of these were built for the Royal Saxon State Railways from 1892 to 1921. They were articulated, and used the Günther-Meyer system of powered trucks; 22 have survived.

Merddin Emrys

The FR Double Fairlie No.10 *Merddin Emrys* was built to combine large haulage capacity with route flexibility. Originally designed by Robert Francis Fairlie and championed by the Ffestiniog Railway in North Wales, Double Fairlie articulated locomotives were able to negotiate tight curves thanks to their flexible steam pipes and pivoting power trucks. Fairlie's patented design was also used in Russia, Mexico, Germany, Canada, Australia, and the United States.

ON JULY 21, 1879, almost 10 years since the first of Robert Fairlie's double-ended articulated locomotives had arrived on the Ffestiniog Railway (FR), *Merddin Emrys* was rolled out of the railroad's Boston Lodge workshops. Designed by G. P. Spooner using Fairlie's principles, No.10 *Merddin Emrys* was the third Double Fairlie to be employed on the FR and it can still be seen there today. The locomotive could comfortably haul 80-ton loads uphill from harbor at Porthmadog to the slate quarries at Blaenau Ffestiniog 3 miles (21 km) away. Impressively, some of these trains were up to 1,312 ft (400 m) long.

The design featured a double-ended boiler with two separate fireboxes in the center. Unlike conventional steam locomotives that carried their boilers on a rigid frame, the boiler and superstructure of the Double Fairlie were supported at each end by a short-wheelbase power truck, connected by flexible steam hoses. This allowed the trucks to turn into a curve before the main body of the locomotive. It was possible to drive each "end" of the locomotive independently, with the engineer and fireman standing on either side of the firebox.

TOP END

BOTTOM END

Six of the best
Built in 1836, the Ffestiniog Railway used six Double Fairlie 0-4-4-0T locomotives to transport slate from Blaenau Ffestiniog to the sea at Porthmadog, South Wales.

SPECIFICATIONS			
Class	FR Double Fairlie	**In-service period**	1879-present (Merddin Emrys)
Wheel arrangement	0-4-4-0T	**Cylinders**	4
Origin	UK	**Boiler pressure**	160 psi (11.25 kg/sq cm)
Designer/Builder	R.Fairlie/G.P.Spooner/FR	**Drive wheel diameter**	32 in (813 mm)
Number produced	6 (2 of this improved design)	**Top speed**	35 mph (56 km/h)

Steam dome on each boiler barrel

Engineer's cab is split in half by boiler

Coal bunker within water tank

Separate exhausts serve each end

Boiler is double-ended

Four-coupled power truck on each end

F.R.C. No.10 Fairlie Patent

MERDDIN EMRYS

Twin role
The fireman of a Double Fairlie has to contend with twice the amount of work. There are two fire boxes, but only one boiler, and a common water space. Both fireboxes need to be used to maintain working boiler pressure.

EXTERIOR

While it might appear to be a product of Victorian times, today's *Merddin Emrys* is virtually a new locomotive. In 1970 it was extensively rebuilt with a new boiler, which gave it a larger, less traditional look. By 1973 it was converted to burn oil instead of coal, and by 1984 it was in need of another overhaul. Its builders decided to remake *Merddin Emrys* in its original 1879 appearance, but retained its larger superstructure in line with the Ffestiniog Railway's improved loading gauge restrictions. The "new" locomotive emerged in 1988, only to be overhauled again in 2005. *Merddin Emrys* reverted to being a coal burner in 2007.

1. Name plate **2.** Smokebox door **3.** Water tank filler **4.** Number plate on smokebox **5.** Sandpot **6.** Top end whistle **7.** Mechanical lubricator **8.** Reverser **9.** Crosshead **10.** Handbrake **11.** Bottom end coal bunker **12.** Bottom end engineer's-side truck **13.** Crosshead and cylinder **14.** Drive wheel **15.** Speedometer drive **16.** Small whistle **17.** "Norwegian Chopper" coupler

CAB INTERIOR

The large fireboxes in the center of the cab mean that the engine crew have to stand in confined spaces on either side of the footplate with the firebox between them. The engineer has a single reverser and two regulator handles, which allow the necessary amount of steam to be thrust to either power truck, as and when required. The design and position of the handles enable regulators to be opened simultaneously with one hand. The fireman, meanwhile, has two firehole doors, one for each firebox, and two sets of gauges. Coal is carried in bunkers built into the water tanks on the fireman's side.

18. Bottom end firebox **19.** Water gauge **20.** Boiler pressure gauge **21.** Top end manifold shutoff **22.** Coal bunker door **23.** Vacuum ejector, steam brake, and injector **24.** Vacuum release valve **25.** Top end injector and slacker valve **26.** Top end firebox door

Shrinking the World

Railroad building continued rapidly. In the 1870s India built its first locomotive using parts imported from Britain, and in 1872 Japan opened its first railroad. Elsewhere, larger engines were being introduced and the mass production of freight locomotives began. The earlier introduction of passenger steam engines on the narrow-gauge railroad at Ffestiniog, Wales, led to the adoption of other narrower-gauge railroads especially in mountainous regions because they were cheaper to construct and could cope with sharper curves and steeper gradients.

◁ Japan's No. 1, 1871/2

Wheel arrangement	2-4-0T
Cylinders	2
Boiler pressure	140 psi (10 kg/sq cm)
Drive wheel diameter	52 in (1,320 mm)
Top speed	approx. 30 mph (48 km/h)

Built in the UK by the Vulcan Foundry in 1871, No. 1 was the first steam locomotive to operate on Japan's inaugural public railroad, from Tokyo to Yokohama, which opened in 1872. From 1880 it went to work on other Japanese railroads before retiring in 1930. It is now on display at the Saitama Railway Museum.

▷ V&TRR No. 20 *Tahoe*, 1875

Wheel arrangement	2-6-0
Cylinders	2
Boiler pressure	130 psi (9.14 kg/sq cm)
Drive wheel diameter	48 in (1,220 mm)
Top speed	approx. 30 mph (48 km/h)

Built by the Baldwin Locomotive Works, Philadelphia, in 1875, No. 20 *Tahoe* worked on the Virginia & Truckee Railroad in Nevada, US, until 1926. The 37.4-ton (38-metric-ton) locomotive was temporarily brought out of retirement during WWII. It has since been restored and is now on display at the Railroad Museum of Pennsylvania in Strasburg.

TALKING POINT

Prince of Wales's Coach

Constructed at the Agra Workshops of the 3-ft 3-in- (1-m-) gauge Rajputana Malwa Railway in 1875, this elegant coach was specially built for the then Prince of Wales (later King Edward VII) for his visit to India in 1877. The prince traveled to India for the Royal Durbar, which celebrated the coronation of his mother Queen Victoria as Empress of India. With all of its original fixtures intact, this coach is now on display at the National Rail Museum, New Delhi.

Royal transportation This unique, four-wheel coach has balconies at each end with seating for four armed guards. The car has sunshades on both sides and is decorated with emblems of the British Crown.

△ Indian F Class, 1874

Wheel arrangement	0-6-0
Cylinders	2
Boiler pressure	approx. 140 psi (10 kg/sq cm)
Drive wheel diameter	approx. 57 in (1,480 mm)
Top speed	approx. 30 mph (48 km/h)

Derived from the British-built 3-ft 3-in- (1-m-) gauge F Class mixed traffic locomotives introduced in 1874, F1 Class No. 734 was the first locomotive to be assembled in India, using imported parts. It worked on the Rajputana Malwa Railway from 1895, and is now an exhibit at the National Rail Museum, New Delhi.

▷ GWR 2800 Class, 1903/1905

Wheel arrangement 2-8-0

Cylinders 2

Boiler pressure 225 psi (16.82 kg/sq cm)

Drive wheel diameter 55½ in (1,410 mm)

Top speed approx. 50 mph (80 km/h)

Eighty-four of these heavy freight locomotives, designed by George Churchward, were built at the Great Western Railway's Swindon Works between 1903 and 1919. Most were in service until the early 1960s.

◁ GWR Steam Railmotor, 1903

Wheel arrangement 0-4-0 + 4-wheel unpowered truck

Cylinders 2

Boiler pressure 160 psi (11.25 kg/sq cm)

Drive wheel diameter 48 in (1,220 mm)

Top speed approx. 30 mph (48 km/h)

Built by the Great Western Railway, these self-propelled cars were equipped with a steam-powered truck and a vertical boiler at one end, and an engineer's compartment at both ends. The railmotors operated suburban passenger services in London, and on rural branch lines in England and Wales. A re-creation was completed by the Great Western Society in 2011 using an original body and a new power truck.

△ LTSR Class 79, 1909

Wheel arrangement 4-4-2T

Cylinders 2

Boiler pressure 170 psi (11.95 kg/sq cm)

Drive wheel diameter 78 in (1,980 mm)

Top speed approx. 65 mph (105 km/h)

Four of these suburban tank engines, designed by Thomas Whitelegg, were built for the London, Tilbury & Southend Railway's commuter services from Fenchurch Street station in 1909. Retired in 1956, *Thundersley* is now part of the UK's national collection.

◁ GWR Iron Mink Covered Wagon, 1900

Type 4-wheel

Weight 11.2 tons (10.16 metric tons)

Construction iron

Railroad Great Western Railway

More than 4,000 of these covered wagons were built by the Great Western Railway from 1886 to 1902. Ventilated and refrigerated versions carried meat, fish, and fruit. Truck versions weighing 33.6 tons (30.5 metric tons) were built between 1902 and 1911.

△ The Royal Daylight Tank Wagon, 1912

Type 4-wheel

Weight 15.7 tons (14.2 metric tons)

Construction iron

Railroad private owner

Built for the Anglo-American Oil Co. by Hurst Nelson of Motherwell, UK, this private-owner tank wagon carried imported American lamp oil branded as Royal Daylight. It is now displayed at Didcot Railway Centre.

GWR Auto Trailer No. 92

Great Western Railway's Auto Trailer No. 92, built at Swindon Works in the UK in 1912 and now based at Didcot Railway Centre, is a unique survivor of one of the earliest types of GWR "auto coach." It is essentially a passenger car with a built-in driving compartment at one end, with controls that link to the steam railmotor to which it is coupled as a two-car unit. The ensemble can therefore be driven in either direction without the need for the locomotive to "run round" when it has reached its destination.

RESTORED TO ITS ORIGINAL GWR Crimson Lake livery, the 70-seater Auto Trailer No. 92 is the non-powered, trailing half of the Great Western Society's Railmotor & Trailer set. The powered half is the railmotor itself (No. 93, pictured above), a near-identical timber-bodied vehicle that has its own built-in, vertical-boilered steam engine, and seating for 50 passengers. The two vehicles ran coupled together as a "steam multiple unit"—the ancestor of today's modern multiple-unit trains—on GWR's branch lines and on their main lines as a "stopping" passenger train.

When operating railmotor first, the engineer and fireman work in the engine compartment. When traveling auto trailer first, the fireman remains with the engine operating the valve gear and injectors and feeding the fire, while the engineer moves to a compartment at the front of the auto trailer. From there he has command of the unit's basic controls, which are connected to the engine by a series of interacting rods, linkages, pipes, or chains. He can also sound a warning bell on the front of the coach.

FRONT VIEW

REAR VIEW

SPECIFICATIONS FOR RAILMOTOR			
Class	Railmotor	In-service period	1912–57 (No. 93)
Wheel arrangement/cylinder	0-4-0 + 4-wheel bogie	Cylinders	2
Origin	UK	Boiler pressure	160 psi (11.25 kg/sq cm)
Designer/builder	George J. Churchward	Drive wheel diameter	48 in (1,220 mm)
Number produced	18 railmotors	Top speed	approx. 30 mph (48 mph)

SPECIFICATIONS FOR AUTO TRAILER NO. 92	
Origin	UK
In-service	1912–57
Coaches	1 (couples with railmotor)
Passenger capacity	70 seats (plus 50 in railmotor)
Route	Great Western Railway routes

Corridor connection to railmotor, which contains the engine

Luggage compartment at rear of auto trailer

Smoking saloon has capacity for 30 passengers

Central entrance vestibule with retractable steps

Non-smoking saloon seats 40 passengers

Driving compartment is used when the auto trailer is in front

Driving compartment
From his forward-facing driving compartment at one end of the auto trailer, the engineer has command of a regulator lever and vacuum brake, which are connected to the steam railmotor, and a bell to signal to the conductor and the fireman.

Great Western cities
The garter design for GWR's coat of arms, which includes the heraldic shields of the cities of London and Bristol, was adopted from 1870 and displayed extensively throughout their system.

EXTERIOR

In the early and later years of the GWR, its coaches were all finished in a brown and cream livery, but in 1912–22 the railroad standardized on a dark red called Crimson Lake. Completed in 1912, the auto trailer was finished in this Crimson Lake livery with gold lining—some 1,200 ft (366 m) of it—and GWR insignia. The recent restoration project, completed in 2012, has returned the auto trailer to these original colors.

1. Car number **2.** Engineer's warning bell **3.** Coat of arms of the City of Bristol **4.** Destination board attached to side of car **5.** Sign on luggage compartment door **6.** Fold-down passenger steps into car **7.** Brass door handle to passenger compartment **8.** Pressure gauge on gas tank **9.** Secondary suspension of transverse leaf springs **10.** Part of truck (wheel unit) **11.** Gas tank for car lighting **12.** Rear buffer

DRIVING COMPARTMENT

The spacious interior of the driving compartment gives the engineer command of the train's basic controls. It also includes a fold-down seat, but this is rarely used, since the engineer has to stand to reach and operate the regulator. Communication between the engineer, fireman, and guard is via an electric (battery-powered) bell, and a series of simple bell codes: one ring for "start," two for "stop," and three for "brakes off." For the engineer's comfort, there is a steam-heat radiator, and there are windshield wipers, too—but 1912 technology did not extend to an electric motor to run them, so manual operation is necessary.

13. Cab interior, with regulator lever above central window to allow engineer to control the steam railmotor from the auto trailer **14.** Vacuum gauge **15.** Lever to open sandbox **16.** Bell to signal to other members of train crew **17.** Vacuum brake control **18.** Foot treadle to sound exterior warning bell

CAR INTERIOR

Restored by craftsmen at the Llangollen Railway in North Wales, the seating in No. 92's two passenger saloons is authentically upholstered in GWR-style diamond-pattern brown moquette. Some of the seats featuring "flip over" backs, which allow passengers to face the direction of travel, were recovered from a derelict tramcar in Adelaide, Australia.

19. Overview of car interior **20.** Replica of original gas light fixture, now powered by electricity **21.** Window shade **22.** Wooden, hand-carved corbel **23.** Electric light switches (a modern addition) **24.** Hand strap suspended from ceiling with decorative metal brackets **25.** Armrest between seats **26.** Part of heater under seats, fed with steam from the railmotor boiler **27.** Metal seat leg **28.** Smoking saloon sign on glass window **29.** Match striker in smoking compartment **30.** Leather strap to open and close window **31.** Emergency pull chain **32.** Decorative brass handles on door leading to car **33.** Ticket rack in guard's vestibule between passenger saloons **34.** Lever for releasing exterior fold-down steps **35.** Twin luggage doors **36.** Luggage door locking mechanism **37.** Wicket gate at end of car leads to railmotor compartment

Continental Glamour

Railroads were now integral to the lives of Europeans. Trains carried vast quantities of raw materials and finished goods as well as large numbers of passengers. Traveling times between European cities had been cut significantly thanks to improvements in tracks and signaling, and also to modern coaches and powerful locomotives capable of sustaining higher speeds for greater lengths of time. New technology led the way as superheated and compound engines rolled off the production lines in ever greater numbers, while the US-influenced 4-6-2 "Pacific" type also started to make an appearance.

◁ **Nord Compound, 1907**

Wheel arrangement	4-6-0
Cylinders	4 (compound)
Boiler pressure	232 psi (16.3 kg/sq cm)
Drive wheel diameter	69 in (1,750 mm)
Top speed	approx. 70 mph (112 km/h)

French engineer Alfred de Glehn designed these compound express locomotives. Built for railroads in France and abroad, some remained in service until the 1960s.

△ **Bavarian Class S3/6, 1908**

Wheel arrangement	4-6-0
Cylinders	4 (compound)
Boiler pressure	213 psi (15 kg/sq cm)
Drive wheel diameter	73$\frac{1}{2}$ in (1,870 mm)
Top speed	approx. 75 mph (120 km/h)

Designed by the German company Maffei, a total of 159 of these express locomotives were built over a period of nearly 25 years—89 for the Royal Bavarian State Railways and 70 (known as Class 18.4-5) for the Deutsche Reichsbahn—between 1908 and 1931.

TALKING POINT

1895 Paris Crash

On the afternoon of October 22, 1895, an express train from Granville, hauling three baggage cars, a mail car, and six passenger cars, approached the Montparnasse terminus, Paris. The train was traveling too fast, its air brake failed, and it crashed through the buffer stop at 30 mph (48 km/h), then traveled across the station concourse, through the station wall, and down to the street. A woman pedestrian was killed, but amazingly, there were no fatalities on the train.

The infamous accident Locomotive No. 721 lies upended on its nose after crashing through the 2-ft- (60-cm-) thick wall of the terminus and falling 33 ft (10 m) onto the street below.

▽ **Prussian Class P8, 1908**

Wheel arrangement	4-6-0
Cylinders	2
Boiler pressure	170 psi (11.95 kg/sq cm)
Drive wheel diameter	69 in (1,750 mm)
Top speed	68 mph (110 km/h)

One of the most successful European steam locomotive designs, around 3,700 of the Prussian state railroad's superheated Class P8s were built between 1908 and 1926. Designed by Robert Garbe, they were built in several different German factories.

△ **SJ B Class, 1909**

Wheel arrangement	4-6-0
Cylinders	2
Boiler pressure	171 psi (12 kg/sq cm)
Drive wheel diameter	69 in (1,750 mm)
Top speed	approx. 65 mph (105 km/h)

The Swedish state railroad (Statens Järnvägar, or SJ) built 96 of these powerful superheated locomotives between 1909 and 1920. Three more were made in 1944. The engines were used to haul express passenger and freight trains.

◁ **PO Pacific, 1910**

Wheel arrangement	4-6-2
Cylinders	2
Boiler pressure	approx. 200 psi (14.06 kg/sq cm)
Drive wheel diameter	67 in (1,702 mm)
Top speed	56 mph (90 km/h)

Built for the Paris à Orléans Railway, these express locomotives were the first "Pacific" type in mainland Europe. Fifty were built in the US by the American Locomotive Co. (ALCO).

△ **FS Class 740, 1911**

Wheel arrangement	2-8-0
Cylinders	2
Boiler pressure	171 psi (12 kg/sq cm)
Drive wheel diameter	55 in (1,400 mm)
Top speed	approx. 56 mph (90 km/h)

A total of 470 of these mixed-traffic engines were built for the Italian state railroad (Ferrovie dello Stato, or FS) between 1911 and 1923, some remaining in service until the 1970s. No. 740.423 has been restored to operational condition in Sardinia and is occasionally used on charter trains.

△ **Prussian Class T18, 1912**

Wheel arrangement	4-6-4T
Cylinders	2
Boiler pressure	170 psi (11.95 kg/sq cm)
Drive wheel diameter	65 in (1,650 mm)
Top speed	approx. 62 mph (100 km/h)

The last tank locomotive designed for the Prussian state railroad, 534 Class T18s were built between 1912 and 1927. Some were still in service in the 1970s with Deutsche Bundesbahn in West Germany and Deutsche Reichsbahn in East Germany.

Fulgence Bienvenüe
1852–1936

French civil engineer Fulgence Bienvenüe was the creator of the Paris Métro, a network that revolutionized the daily lives of Parisians. His extraordinary achievement followed an inauspicious beginning to his railroad career; in 1881 he had lost his left arm in a construction accident while working on his first rail project in Normandy, France. However, this did not deter him from pursuing his engineering ambitions, and after moving to Paris in 1886, he became chief engineer for the Métro and supervised its development over the next 35 years. In addition to the Métro, Bienvenüe also managed engineering projects for the Parisian highway, lighting, and cleaning departments.

FATHER OF THE METRO

With Paris hosting the Universal Exhibition in 1900, the city's Municipal Council asked Bienvenüe to draw up plans for a narrow-gauge metro network for electric trains. The project started on October 4, 1898 and the first Métro line (Line 1, Porte de Vincennes to Porte Maillot) opened to passengers on July 19, 1900, in time for the exhibition.

The speed and efficiency of this new urban transit system impressed Parisians so much that the council granted Bienvenüe the job of extending and building a full underground network. Progress was swift. Within five years Lines 2 and 3, which stretched for 26 miles (42 km), were completed despite a number of unforeseen setbacks, including a fire at Couronnes in 1903 in which 84 people died. When Line 4 was tunneled under the Seine River (1904–10), the construction techniques used were hailed as masterstrokes of civil engineering. By the eve of World War I, the Paris Métro was largely complete.

In 1933 the Avenue du Maine station was renamed Bienvenüe in honor of the "father of the Paris Métro." Nowadays, with some 1.5 billion journeys made on the Métro each year, the network is an integral part of the city.

Early Paris Métro
Three Métro lines (3, 7, and 8) cross one another beneath the Place de l'Opéra. The enormous construction effort to build the Métro saw the streets of central Paris torn up, much to the alarm of Parisians.

Honoring history
A Sprague-Thomson electric train arrives at
Place de la Bastille Métro station on Line 1 in
1912. Paris Métro stations are named after
significant events, places, and people from
French history.

H&BT Caboose No. 16

Built by the Pennsylvania Railroad in 1913, this wooden, four-wheeled caboose, or cabin car, saw service on the railroad's Middle Division between Harrisburg and Altoona before being sold to the Huntingdon & Broad Top Mountain Railroad & Coal Company (H&BT). Known as bobbers, these four-wheeled cabooses were attached to the rear end of a freight train, serving as an office, lookout, and home for the crew during trips.

WIDELY USED ON NORTH AMERICAN railways from the 1870s through to the 1930s, bobbers got their name from train crews for their bumpy and occasionally unstable riding conditions. The cupola on the roof offered all-around visibility for conductors, allowing them to watch the freight cars during their journeys. Originally numbered No. 478396, this caboose was built at the Pennsylvania Railroad's Car Shops in Altoona and remained in service until 1940, when it was sold to the Huntingdon & Broad Top Mountain Railroad & Coal Company. It was then renumbered No. 16 on this coal-carrying short line in south-central Pennsylvania and was one of the last wooden-bodied, four-wheeled bobbers in service in the US when the railroad closed in 1954. Saved from the scrapyard, it then had several owners before it was donated to the Railroad Museum of Pennsylvania in 1998. Here it was expertly restored and is currently on display in H&BT red livery.

REAR VIEW

FRONT VIEW

H.&B.T.

Bright bobber
Restored caboose No. 16 now carries the initials of the Huntingdon & Broad Top Mountain Railway (H&BT), which originally opened in 1855 to serve coal mines in Pennsylvania.

SPECIFICATIONS			
Type	Caboose (cabin car)	In-service period	1913-54
Origin	United States	Passenger capacity	1 conductor, crew's quarters
Designer/builder	Altoona Car Shops	Weight	14 tons (12.7 metric tons)
Number produced	Not known	Railroad	Pennsylvania Railroad/H&BT

Chimney for crew stove

Cupola serves as observation position for brakeman

Marker lamp signaled end of train

Ladder for access to roof

Veranda allows easy access to brake controls and roof

Handbrake allows conductor to slow train on downgrades

Buckeye coupling could be operated manually using the coupling opener arm

Mounting steps made of steel, with wooden treads

H.&B.T.
16

Sax Wt 29200 12 40

Crew's caboose
North American cabooses traditionally had
a veranda at each end, which was reached
by steps from ground level. A steel ladder
enabled freight train crew to access
the roof in order to clean the windows
of the lookout cupola.

EXTERIOR

The caboose had a cupola from which the conductor kept a lookout for overheating axles, or hotboxes, as well as shifting cargo and damage to the train. The buckeye coupling could be manually operated; use of this type of coupling and air brakes was made mandatory by Congress in 1893, significantly reducing the number of railroad accidents and workers killed or injured during coupling operations.

1. Caboose number painted on side   **2.** Coupling link
3. Coupling opener arm   **4.** Steps up to end platform
5. Marker lamp   **6.** Retaining valve to keep air brakes applied on long descending grades   **7.** Whistle   **8.** Chimney
9. Windows in observation level   **10.** Wheel unit   **11.** Open journal box showing bearing   **12.** Brake wheel on platform

INTERIOR

The caboose's cozy interior was an office and a temporary home for the locomotive crew and conductor. Raised seats allowed views through the roof cupola, and a coal-fired stove bolted to a steel plate on the floor kept the crew warm at night and provided cooking facilities. Surrounded by protective steel plates, the stove was equipped with safety features such as a double-latched door to prevent hot coals from spilling out, and a lip on the top to stop pans and pots from sliding off when the train was in motion.

13. Interior of caboose **14.** Window latch **15.** Air brake pressure gauge **16.** Seats on observation level **17.** Oil lamp **18.** Air controls on coal stove **19.** Coal stove **20.** Sink unit

Rapid Development

With railroads now well established, this period saw rapid developments in the design of both passenger and freight locomotives around the world. Mass production of heavy freight engines reached new heights, with more than 1,000 of the Prussian state railroad's Class G8, along with another 5,000 of the later Class G8.1, being built over the following years. However, the world record for the most numerous class of locomotive goes to the Russian E Class, of which around 11,000 were built.

▷ **Austrian Gölsdorf Class 170, 1897**

Wheel arrangement	2-8-0
Cylinders	2 (compound)
Boiler pressure	185 psi (13 kg/sq cm)
Drive wheel diameter	49½ in (1,260 mm)
Top speed	approx. 37 mph (60 km/h)

Designed by Karl Gölsdorf for the Imperial Royal Austrian State Railways, the Class 170 freight locomotives were the first to be equipped with radially sliding coupled axles, known as Gölsdorf axles.

△ **Prussian Class G8, 1902**

Wheel arrangement	0-8-0
Cylinders	2
Boiler pressure	170 psi (11.95 kg/sq cm)
Drive wheel diameter	53 in (1,350 mm)
Top speed	approx. 35 mph (56 km/h)

More than 1,000 of these superheated freight locomotives were built in Germany for the Prussian state railroads. After WWI, hundreds were given to Germany's enemies as reparations. Some saw service during the building of the Baghdad Railway in Turkey in 1916.

△ **PRR Class E7, 1902**

Wheel arrangement	4-4-2
Cylinders	2
Boiler pressure	205 psi (14.4 kg/sq cm)
Drive wheel diameter	78½ in (2,000 mm)
Top speed	approx. 80 mph (129 km/h)

The original Class E7 No. 7002 was built at the Pennsylvania Railroad's Altoona Works. It was once claimed to be the world's fastest steam engine, supposedly reaching 127 mph (204 km/h), but this is disputed. First numbered 8063, this locomotive was renumbered after the first 7002 was scrapped and is now in the Railroad Museum of Pennsylvania.

◁ **Indian Class EM, 1907**

Wheel arrangement	4-4-2
Cylinders	2
Boiler pressure	190 psi (13.4 kg/sq cm)
Drive wheel diameter	78 in (1,980 mm)
Top speed	approx. 60 mph (96 km/h)

Originally built as a 4-4-0 by the North British Locomotive Co. for the Great Indian Peninsula Railway, the Class EM remained in service until the late 1970s. EM No. 922 was rebuilt in 1941 by the Mughalpura workshops.

△ VGN Class SA, 1910

Wheel arrangement	0-8-0
Cylinders	2
Boiler pressure	200 psi (14.06 kg/sq cm)
Drive wheel diameter	51 in (1,295 mm)
Top speed	approx. 10 mph (16 km/h)

One of only five Class SA switcher locomotives built, Nos. 1, 2, and 3 were made at American Locomotive Co. (ALCO), Nos. 4 and 5 by Baldwin Locomotive Works. No. 4 (shown) retired in 1957 as the last steam locomotive on the Virginian Railway.

▽ Austrian Gölsdorf Class 310, 1911

Wheel arrangement	2-6-4
Cylinders	4 (compound)
Boiler pressure	220 psi (15.46 kg/sq cm)
Drive wheel diameter	84¼ in (2,140 mm)
Top speed	approx. 62 mph (100 km/h)

Designed by Karl Gölsdorf, 90 of the Class 310 four-cylinder compound express locomotives were built for the Imperial Royal Austrian State Railways from 1911 to 1916. This was one of the most elegant locomotives of the period.

△ Russian E Class, 1912

Wheel arrangement	0-10-0
Cylinders	2
Boiler pressure	170 psi (11.95 kg/sq cm)
Drive wheel diameter	48 in (1,220 mm)
Top speed	approx. 30 mph (48 km/h)

First built at Lugansk Works in Ukraine, a large number of these heavy freight locomotives were eventually constructed in Russia, as well as in countries such as Czechoslovakia, Germany, Sweden, Hungary, and Poland. There were several subclasses, some of which were equipped with condensing tenders for working in areas where water was scarce.

TECHNOLOGY

Geared Locomotives

US-built, lighter-weight geared steam locomotives such as the Shay, Heisler, and Climax types had wheels driven by reduction gearing. These locomotives were designed for the quick and cheap-to-lay industrial railroads used by logging, sugar cane, mining, and quarrying industry operations where speed was not needed and gradients were often steep.

Heisler 2-truck geared locomotive No. 4 This locomotive, designed by Charles L. Heisler, was built for the Chicago Mill & Lumber Co. in 1918. It was the fastest of this type and is on display at the Railroad Museum of Pennsylvania.

VGN Class SA No. 4

One of only five Class SA 0-8-0 switchers, this powerful locomotive was delivered by the Baldwin Locomotive Works of Eddystone, PA, to the newly formed Virginian Railway (VGN) in August 1910. It marshaled heavy coal trains at the railroad's yards in Virginia and West Virginia until its retirement in 1957, when it was replaced by diesel locomotives. It is currently on display at the Virginia Museum of Transportation in Roanoke, and is the last surviving steam engine of the Virginian Railway.

OPENED IN 1909, THE VIRGINIAN RAILWAY became a highly profitable company by transporting high-quality coal from the mines in West Virginia to its piers at Sewells Point, Norfolk, southwestern Virginia, from where it was it was transferred on to ships. Nicknamed the "Richest Little Railroad in the World," the railroad used some of the world's most powerful steam locomotives to haul its heavy eastbound coal trains up the steeply graded line to Clark's Gap in West Virginia, until this section of the line was electrified in 1925.

Marshaling the long coal trains in Page (named after one of the railroad's founders) and other yards in West Virginia and Virginia, was carried out by powerful 0-8-0 Class SA switchers, of which No. 4 is the only surviving example. Of the five Class SA switchers built, Nos. 1–3 were supplied by ALCO and Nos. 4–5 by the Baldwin Locomotive Works.

FRONT VIEW

REAR VIEW

Baldwin Locomotive Works
Founded by Matthias Baldwin in 1825, the Baldwin Locomotive Works built more than 70,000 engines for railroads around the world. In 1956, production ceased after it lost out on a large order to supply diesels for the Pennsylvania Railroad.

SPECIFICATIONS			
Class	SA	In-service period	1910-57 (No. 4)
Wheel arrangement	0-8-0	Cylinders	2
Origin	United States	Boiler pressure	200 psi (14.06 kg/sq cm)
Designer/builder	Baldwin Locomotive Works	Drive wheel diameter	51 in (1,295 mm)
Number produced	5 Class SA	Top speed	approx. 10 mph (16 km/h)

Tender has a water capacity of 5,000 gallons (18,927 liters)

Coal bunker can hold 10 tons

Rear sand dome sands track behind drive wheels when backing up

Steam dome contains throttle

Front sand dome sands track ahead of drive wheels when going forward

Powerful switcher
A utilitarian machine, Class SA No. 4 weighed in at 81 tons, and had a tractive effort of 45,200 lb (20,52 kg). The rear eight-wheeled tender weighed almost 50 tons when fully loaded.

EXTERIOR

SA No. 4 was built as a utilitarian workhorse able to marshal heavy coal trains at slow speeds in switching yards. It was equipped with knuckle couplings and a Westinghouse air brake, both US standard systems. The two air reservoirs for the brakes were housed between the two cylinders at the front of the locomotive.

1. Engine number on side 2. Headlight 3. Front coupler 4. Valve chmber head cover with metal star detail 5. Builder's plate on side of engine 6. Brass bell on top of engine 7. Whistle attached to steam dome 8. Safety valves 9. Piston rod 10. Crosshead support yoke 11. Drive wheels 12. Drive wheel springs 13. Steps leading to cab 14. Exterior of cab with bright red window frames 15. Tender behind cab 16. Signage displaying tender water capacity 17. Light on tender 18. Handrail around edge of tender

CAB INTERIOR

The engineer and fireman of SA No. 4 worked in a hot and uncomfortable cab. The driver was seated on the right-hand side, where he could see the tracks ahead and control the throttle and air brake. Unlike the many American locomotives equipped with mechanical stokers, the humble switchers had to be manually fed coal from the tender into the firebox by the fireman, using a large shovel.

19. Boiler backhead in cab **20.** Engine and train brake **21.** Steam pressure gauge **22.** Interior of firebox **23.** Air brake gauge
24. Auxiliary controls **25.** Throttle lever (regulator) **26.** Control valves
27. Control pedal **28.** Engineer's seat

The New York Elevated Railway

While London and Paris burrowed underground to meet the demand for a fast and reliable mass transit system, New York chose the overground route. Between 1840 and 1870, the city's population had grown by more than half a million inhabitants. This increase overwhelmed the capacity of its horse-drawn bus and streetcar routes, several of which ran along the main avenues. Although it was considered unsafe and impractical to replace horses with steam engines, two local entrepreneurs, Charles Harvey and Rufus Gilbert, believed their locomotives could run on viaducts built over the streets. They introduced two elevated lines to the west side of Manhattan before financial problems forced the city to take over the project.

Under the Rapid Transit Act of 1875, four lines were constructed, and these would form the heart of the New York Elevated Railway (or the "El," as New Yorkers called it). The routes ran along Second, Third, Sixth, and Ninth avenues, and further lines were added up to 1917.

Although the smoke and noise of steam locomotives had been supplanted by electric traction, by the late 1930s the "El" was considered outdated. The lines were demolished between 1938 and 1955 to make way for the New York subway system.

Passengers ride behind a lightweight, Forney tank locomotive of the Third Avenue Elevated Railroad in 1896, above the wagons and streetcars of Bowery

On Other Gauges

George Stephenson introduced the 4-ft 8½-in (1.43-m) gauge for British railroads in 1830, and before long it became the standard gauge for many railroads around the world. However, there were, and still are, many exceptions. In India, a broader gauge of 5 ft 6 in (1.67 m) was used for many main rail lines, but more lightly laid lines had narrower gauges of 3 ft 3 in (1 m) or, for mountain railroads, only 2 ft (0.61 m). While the standard gauge was usually the norm in mainland Europe and the US, there was also widespread use of narrow gauges in mountainous regions. The most extensive narrow-gauge network in the US was the Denver & Rio Grande Railroad's 3-ft (0.91-m) gauge system in Arizona, Utah, and New Mexico.

▷ NWE Mallet, 1897

Wheel arrangement	0-4-4-0
Cylinders	4
Boiler pressure	200 psi (14 kg/sq cm)
Drive wheel diameter	39½ in (1,000 mm)
Top speed	approx. 18 mph (30 km/h)

This engine was one of 12 powerful articulated steam locomotives built for the German 3ft 3-in- (1-m-) gauge Nordhausen-Wernigerode Railway. Several were lost in WWI but three are now with the NWE's successor, running on the Selke Valley Railway in the Harz Mountains in central Germany.

△ NWR ST, 1904

Wheel arrangement	0-6-2T
Cylinders	2 (inside)
Boiler pressure	150 psi (10.53 kg/sq cm)
Drive wheel diameter	51 in (1,295 mm)
Top speed	approx. 30 mph (48 km/h)

One of the first locomotives built at India's North Western Railway's Mughalpura Workshops, ST No.707 was made from parts supplied by North British Locomotive Co. of Glasgow. Weighing 55 tons, this 5-ft-6-in- (1.67-m-) gauge locomotive was employed for shunting duties. It is now on display at the National Rail Museum, New Delhi.

△ KS Wren Class, 1905

Wheel arrangement	0-4-0
Cylinders	2
Boiler pressure	140 psi (9.84 kg/sq cm)
Drive wheel diameter	20 in (500 mm)
Top speed	approx. 15 mph (24 km/h)

A total of 163 of these narrow-gauge locomotives were built by the British company Kerr Stuart for use on industrial railroads around the world between 1905 and 1930. However, *Jennie* was made in 2008 for the 2-ft (0.61-m) gauge Amerton Railway, Staffordshire, by the Hunslet Engine Co.

▷ Indian SPS, 1903

Wheel arrangement	4-4-0
Cylinders	2 (inside)
Boiler pressure	160 psi (11.25 kg/sq cm)
Drive wheel diameter	78 in (1,980 mm)
Top speed	approx. 50 mph (80 km/h)

A range of standard designs was introduced for India, including the Standard Passenger (SP); when superheating was added, it became the SPS. Mostly built in the UK, some of these engines had extremely long working lives. After Partition in 1947, this one ran on the new Pakistan Railways until the 1980s.

△ **Mh 399, 1906**

Wheel arrangement	0-8+4
Cylinders	2
Boiler pressure	180 psi (12.65 kg/sq cm)
Drive wheel diameter	36 in (910 mm)
Top speed	approx. 25 mph (40 km/h)

Built by Krauss of Linz, this locomotive was made for the Austrian Railways' 2-ft-6-in (0.76-m) narrow-gauge Mariazell Railway. It had rear wheels that are also driven by coupling rods. Seen here is No. 399.06 preserved on the Mariazellerbahn, Austria.

△ **TGR K Class Garratt, 1909**

Wheel arrangement	0-4-0+0-4-0
Cylinders	4
Boiler pressure	195 psi (13.70 kg/sq cm)
Drive wheel diameter	31½ in (800 mm)
Top speed	approx. 25 mph (40 km/h)

The world's first Garratt-type articulated steam locomotive, No. K1 was built by Beyer Peacock & Co. of Manchester, England, for the Tasmanian Government Railway, Australia. It ran on the 2-ft (0.60-m) gauge North East Dundas Tramway. This historic locomotive was returned to Britain in 1947 and now hauls trains on the Welsh Highland Railway.

△ **EIR No. 1354 *Phoenix*, 1907**

Wheel arrangement	0-4-0WT
Cylinders	2 (inside)
Boiler pressure	120 psi (8.44 kg/sq cm)
Drive wheel diameter	36 in (910 mm)
Top speed	approx. 20 mph (32 km/h)

One of five railmotors built in England by Nasmyth Wilson & Company, *Phoenix* was built for the 5-ft-6-in (1.67-m) East Indian Railway in 1907. Later, in 1925, the coaches were removed and *Phoenix* was rebuilt in India as a small shunting engine. It is now on display at the National Rail Museum, New Delhi.

▷ **Lima Class C Shay, 1906**

Wheel arrangement	B-B-B
Cylinders	3
Boiler pressure	200 psi (14.06 kg/sq cm)
Drive wheel diameter	36 in (910 mm)
Top speed	approx. 15 mph (24 km/h)

Designed by US inventor Ephraim Shay, the Class C geared three-truck steam locomotive was first introduced in 1885. This Shay No. 1 was built by the Lima Locomotive & Machine Co. for a standard-gauge logging railroad in Pennsylvania in 1906. It can be seen at the Railroad Museum of Pennsylvania, Strasburg.

Building Great Railroads
Trans-Siberian Railway

Crossing eight time zones, the 5,772-mile (9,289-km) Trans-Siberian Railway is the longest continuous rail line in the world. Extending from the Russian capital, Moscow, to Vladivostok on the Pacific coast, it provides a strategic route connecting Asia with Europe.

Kama River, near Perm
A metal-truss railroad bridge straddles the Kama River in this early color photograph from c.1909-15. The Trans-Siberian crosses numerous major rivers.

BY 1890 THE RUSSIAN EMPIRE stretched east from its European borders, across the Ural Mountains and the vastness of Siberia, to the Pacific coast. While European Russia, west of the Urals, had experienced industrial growth and acquired railroads in the 19th century (the first railroad, opened in 1851, was between Moscow and St. Petersburg), the lands to the east remained virtually untapped. With few roads into the region, the only means of transportation were the mighty Siberian river systems, but these were only navigable for around five months of each year—for the remaining months, they were frozen. A railroad was the key to opening up this vast hinterland.

Construction of the government-funded Trans-Siberian Railway began in 1891 with the blessing of Czar Alexander III and his son, the future Czar Nicholas II. Work began at both ends—Moscow and Vladivostok—with Russian soldiers and convicts employed as railroad navvies. Progress was fast and by 1898 the line stretched 3,222 miles (5,185 km) from Moscow to Irkutsk, near the western shore of Lake Baikal.

Farther east, the line running from Vladivostok to Khabarovsk had already opened in 1897. However, the Amur line running west from Khabarovsk to Chita would not open until much later. Because of difficult terrain in this region, a shortcut linking

Steaming around Lake Baikal
Golden Eagle Trans-Siberian Express is one of the luxury trains that runs the route. Less decadent trips can be taken on a variety of domestic and international services.

RAILROAD EMBLEM

R U S S I A N   F E D E R A T I O N

River crossings On its long route across Russia, the railroad crosses many great rivers, including the world's fifth-longest river, the Ob, at Omsk.

Kirov
Yaroslavl'
Perm'
Yekaterinburg
Moscow
Omsk
Novosibirsk
Krasnoyarsk

BELARUS

3 Moscow
Russia's capital city, Moscow, is shown in 1890, a year before construction on the Trans-Siberian began. Trans-Siberian passenger trains for Vladivostok depart from Yaroslavsky Station, which was opened in 1904.

UKRAINE

2 Construction near Yekaterinburg
The Trans-Siberian Railway was built at the rapid rate of 2 ½ miles (4 km) a day in summer conditions. To reduce costs, lighter rails were used than those standard in Europe.

M O N

KAZAKHSTAN

C H I N A

KEY
● Start/Finish
● Main stations
▦ Main route
▦ Original route 1903
▦ Trans-Mongolian route
▦ Trans-Manchurian route

Vladivostok to Chita via Manchuria was built. However, following conflicts with Japan over Manchurian interests, a route on Russian soil was needed and work on the Amur line began.

Meanwhile, the eastern and western sections of the Trans-Siberian Railway had come to an end on opposite shores of Lake Baikal—at 5,387 ft (1,652 m) the deepest freshwater lake in the world. A train ferry, the icebreaker SS *Baikal*, was launched in 1899 to carry whole trains across the lake. It could carry up to 24 train cars and a locomotive. The ferry service became obsolete in 1905 when the Circum-Baikal Line opened around the rocky western shores of Lake Baikal; its 33 tunnels and 200 bridges were built by convicts and political prisoners at great cost to the state.

The Khabarovsk Bridge over the Amur River was built in 1913 and, with the Amur section completed in 1916, the entire line was opened.

Siberian landscape
Full electrification of the Trans-Siberian was completed in 2002. This earlier passenger train hauled by three diesel-electric locomotives heads through the empty landscape.

KEY FACTS

DATES
1891 Building begins from Vladivostok (east) and Moscow (west) toward the center
1903 Original route via Manchuria is completed
1904 Circum-Baikal around Lake Baikal is finished
1916 Final route is completed and line opens

TRAINS
Train No. 002 *Rossiya* travels eastbound Moscow–Vladivostok; No. 001 runs westbound. A range of domestic Russian trains or direct international trains run from Moscow to Ulan Bator, Mongolia; Beijing, China; and Pyongyang, North Korea. Trains are Russian or Chinese rolling stock, depending on final destination. Luxury trains such as the steam-hauled *Golden Eagle* and *Czar's Gold* also run.

JOURNEY
Moscow to Vladivostok
5,772 miles (9,289 km); 6 days 4 hours, Train No. 002M

RAILROAD
Gauge Broad 4 ft 11 ⅚ in (1.52 m)
Tunnels 33 on Circum-Baikal section; longest passenger tunnel Tarmanchukan, 1.4 miles (2.2 km)
Bridges Track crosses 16 major rivers, including the Volga, Ob, Yenisey, and Oka; the Khabarovsk Bridge over the Amur is longest at 8,500 ft (2,590 m)
Highest point 3,412 ft (1,040 m) at the Yablonovy Mountain pass near Chita
Lowest temperature -80°F (-62°C) between Mogocha and Skovordino on the Amur section

HIGH-SPEED CONSTRUCTION

An amazing feat of human effort, the Trans-Siberian Railway was built by thousands of Russian soldiers, as well as convicts and political prisoners serving sentences of hard labor. After 25 years of construction, its completion fulfilled the dreams of Russia's last czar.

5 6 Circum-Baikal Railway
The Trans-Siberian Railway was later built around Lake Baikal, the world's deepest lake. The track follows the lake's western shoreline, but much of it runs through tunnels.

4 SS *Baikal*
Initially Trans-Siberian trains crossed Lake Baikal on the icebreaker rail ferry SS *Baikal*, the parts of which were built in England and assembled in Russia. It could traverse the lake through ice 3 ft (91 cm) thick.

The Amur line The Chita–Khabarovsk section was completed in late 1916, and was built over very difficult terrain.

7 Vladivostok Station 1893/94
Construction of the Trans-Siberian Railway in the east began in 1891 in the historic port of Vladivostok.

Trans-Manchurian Railway Coaches and freight cars change trucks to operate on the Chinese standard gauge.

Trans-Mongolian Railway Opened in 1955, the Mongolian track has the same broad gauge as Russian rail.

1 Ussuri section
Convict labor was used to construct the section from Vladivostok to Khabarovsk, which was completed in 1897.

Chinese Eastern Line Opened in 1903, this line provided a shortcut to Chita, but conflict with Japan made a route on Russian soil necessary.

Irkutsk • Ulan-Ude • Chita • Khabarovsk • Harbin • Vladivostok • Beijing • GOLIA

N 0 300 600 miles
0 300 600 900 km

Competition From the New Electrics

While steam traction was enjoying its heyday in the late 19th and early 20th centuries, other forms of faster and cleaner rail transportation were being developed. Electric streetcars first started running in Europe and the US during the 1880s, and the technology began to appear on railroads by the early 20th century. Using a mixture of either third-rail or overhead catenary power supplies, electric traction had been introduced on many city commuter lines in the UK and the US by the outbreak of World War I. With their fast acceleration these trains were ideal for lines with high-density traffic; they also eliminated the problem of pollution in urban areas and in tunnels. In the US the electrification of the 2³/₄-mile (4.23-km) Cascade Tunnel in Washington State in 1909 was an early example of clean electric locomotives replacing the asphyxiating fumes of steam engines in confined spaces.

△ Budapest Metro car, 1896

Wheel arrangement 2 x 4-wheel powered trucks	
Power supply 300 V DC, overhead supply	
Power rating 28 hp (20.59 kW) per engine	
Top speed approx. 30 mph (48 km/h)	

Equipped with two Siemens & Halske traction motors, 20 of these electric, double-ended subway cars were built for Continental Europe's first electric underground railroad, which opened in Budapest, Hungary in 1896. Plans for extending the metro with two extra routes were made in 1895, but the lines only opened more than 70 years later, in 1970 and 1976. Following retirement in the early 1970s, car No. 18 was preserved and is on display at the Seashore Trolley Museum, Kennebunkport.

◁ NER gasoline-electric autocar, 1903

Wheel arrangement 2 x 4-wheel trucks (1 powered)	
Transmission 2 traction motors	
Engine gasoline	
Total power output 80 hp (59.6 kW)	
Top speed approx. 36 mph (58 km/h)	

Two of these gasoline-electric railcars were built in 1903 in England at the North Eastern Railway's York Works. The original Wolsey four-cylinder engine that drove generators to power the two electric traction motors was replaced by a six-cylinder 225 hp (168 kW) engine in 1923. The railcars had been withdrawn by 1931. One is being restored at the Embsay & Bolton Abbey Steam Railway in Yorkshire.

▷ Drehstrom-Triebwagen, 1903

Wheel arrangement 2 x 6-wheel trucks, outer axles motorized	
Power supply 6-14 kV DC (25-50 Hz)	
Power rating 1,475 hp (1,100 kW)	
Top speed 130 mph (210 km/h)	

Built by Siemens & Halske and AEG of Germany and equipped with three-phase induction motors, two prototype high-speed Drehstrom-Triebwagen railcars were tested on the Prussian military railroad south of Berlin in 1903. Taking overhead power from a triple catenary, the AEG-built railcar reached 130 mph (210 km/h) between Zossen and Marienfelde on October 28, 1903, a world rail-speed record not broken until 1931.

▷ NER electric locomotive, 1905

Wheel arrangement Bo-Bo	
Power supply 600-630 V DC, third-rail or catenary	
Power rating 640 hp (477 kW)	
Top speed approx. 27 mph (43 km/h)	

Drawing power from either a third rail or an overhead catenary, two of these locomotives were built by British Thomson-Houston for the North Eastern Railway in 1903-04 but were not operational until 1905, when the line was electrified. They worked on a steeply graded freight line to a dockside in Newcastle-upon-Tyne until 1964. One is preserved at the Locomotion Museum in Shildon, County Durham.

NORTH

Ticketing on the Railways

Early railroad companies issued tickets to passengers on handwritten pieces of paper. This was time-consuming and open to fraud by unscrupulous ticket clerks. Invented by Thomas Edmondson, an English station master, the Edmondson train ticket system was introduced in 1842. Using preprinted, durable cards was not only a faster means of issuing tickets, but they were also given unique serial numbers that had to be accounted for by booking clerks each day. Ticket inspectors at stations and on trains punched holes in the tickets to prevent reuse.

Ticket punch Featuring a decorative, three-pointed spike, this silver ticket punch was made by the Bonney-Vehslage Tool Co. for the Baltimore & Ohio Railroad in 1906.

Punch hole

△ **B&O Bo Switcher, 1895**

Wheel arrangement	Bo (0-4-0)
Power supply	approx. 450 V, catenary
Power rating	approx. 15 hp (11.2 kW)
Top speed	approx. 10 mph (16 km/h)

Opened in 1860, the Baltimore & Ohio Railroad's network of railroads serving waterfront warehouses at Fells Point in Baltimore was originally horse-drawn. Overhead streetcar power lines were introduced in 1896 with small electric switchers, like this No.10 built by General Electric in 1909, taking over from horsepower.

▷ **Schynige Platte Class He2/2, 1910**

Wheel arrangement	0-4-0
Power supply	1,500 V DC, overhead catenary
Power rating	295 hp (220 kW)
Top speed	approx. 5 mph (8 km/h)

The 2-ft 7½-in- (0.8-m-) gauge Schynige Platte Railway in the Swiss Bernese Oberland opened using steam power in 1893. This steeply graded mountain rack railroad was electrified in 1914. Four of the original electric engines built by the Swiss Locomotive & Machine Works and Brown Boveri still operate on the railroad.

EASTERN

1914–1939

STEAM'S ZENITH

The New Empire State Express Passing West Point in the Highlands of the Hudson on the New York Central System

New York Central System

STEAM'S ZENITH

In 1914 the world was plunged into a terrible conflict. World War I lasted until 1918, and during the four years of hostilities, railroads played a key role. The ability to move troops, munitions, and supplies by train assumed new importance; in many countries, full-sized locomotives were specially built for the military, while narrow-gauge railroads were created to serve the war effort. The latter were designed to be laid easily and to run close to the front lines.

△ **Ticket for the *Royal Blue*, 1935**
Recalling the glamour of the original *Royal Blue* train, B&O Railroad marketed the revamped service as elegant and luxurious.

At the end of the war, maps were redrawn, and many new or recreated countries found themselves inheriting existing rail systems, which they adapted to meet particular demands inside their new borders. In Germany, a postwar reorganization brought its railroads together to create the Deutsche Reichsbahn. In Great Britain, the government merged the private rail companies to form what became known as the "Big Four."

A desire for progress and increasing rivalry (as well as competition from cars and airplanes) combined to give rise to a new age of speed and streamlining. As the Art Deco visual style took hold across the world, new, futuristic-looking trains were launched. Railroads competed with each other not only by offering greater speed and comfort, but also through clever marketing. Toward the end of the period, in July 1938, Britain's *Mallard* snatched the steam speed record from a German locomotive by reportedly reaching 126 mph (203 km/h)—a figure that has never been officially beaten.

Yet as steam neared its streamlined zenith, the push for speed and modernity created a new breed alongside the giants of steam and new lightweight diesel trains began to appear in North America and Europe during the 1930s.

" There is **more poetry** in the rush of a **single railroad train** across the continent than in all **the gory story of Troy**"

JOAQUIN MILLER, AMERICAN POET

◁ **The new *Empire State Express* poster** by Leslie Ragan advertises America's burgeoning railroad tourism in the 1930s

Key Events

▷ **1914** Outbreak of World War I. Railroads prove to be essential for the transportation of troops and supplies.

▷ **1915** In Germany, Leipzig's main station is completed—the world's largest station measured by floor area.

▷ **1916** The final section of the Trans-Siberian Railway is opened.

▷ **1917** The Trans-Australian Railway is finished. Its route includes the world's longest stretch of straight track at nearly 300 miles (483 km).

▷ **1920** Germany's railroads come under the new Deutsche Reichsbahn.

▷ **1931** Germany's gasoline-powered Schienenzeppelin reaches 143 mph (230 km/h), setting a rail speed record.

▷ **1935** The first section of the Moscow Metro opens.

▷ **1934** Sir Nigel Gresley's steam locomotive *Flying Scotsman* records a speed of 100 mph (161 km/h).

▷ **1936** A German "Leipzig" diesel railcar travels at 127 mph (205 km/h), a record for diesel traction.

▷ **1938** France's railroads are brought together as the Société Nationale des Chemins de fer Français (SNCF).

▷ **1938** Sir Nigel Gresley's *Mallard* hits 126 mph (203 km/h)—a steam speed record that still stands today.

△ **Record-breaking *Mallard***
Mallard and the dynamometer car stand at Barkston on July 3, 1938, poised for the run that will earn the locomotive a world speed record for steam.

Locomotives for World War I

Following the outbreak of World War I, the Railway Operating Division (ROD) of the British Royal Engineers was formed in 1915 to operate railroads in the European and Middle East theaters of war. The British network of narrow-gauge trench railroads was operated by the War Department Light Railways, while the French standardized portable 1-ft 11¾-in- (0.60-m-) gauge, military Decauville equipment to supply the Western Front. The Germans used a similar system for their trench railroads. The entry of the US into the war in 1917 saw many US-built locomotives shipped across the Atlantic for service in France. President Wilson nationalized the US railroads from January 1, 1918 to March 1, 1920 to aid the war effort.

△ GWR Dean Goods, 1883

Wheel arrangement	0-6-0
Cylinders	2 (inside)
Boiler pressure	180 psi (12.65 kg/sq cm)
Drive wheel diameter	61¾ in (1,570 mm)
Top speed	approx. 45 mph (72 km/h)

Designed by William Dean, 260 of these standard-gauge freight locomotives were built at the Great Western Railway's Swindon Works between 1883 and 1899. In 1917 the Railway Operating Division commandeered 62 of them to operate supply trains in northern France. Some also served in France during WWII.

▽ Henschel Meter-gauge, 1914

Wheel arrangement	0-6-0T
Cylinders	2
Boiler pressure	200 psi (14 kg/sq cm)
Drive wheel diameter	31½ in (800 mm)
Top speed	approx. 18 mph (29 km/h)

Built by the German company Henschel in 1914, two of these 3-ft 3-in- (1-m-) gauge locomotives were originally supplied to the Army Technical Research Institute. They were later transferred to the Nordhausen-Wernigerode Railway in the Harz Mountains in central Germany, where they hauled trains carrying standard-gauge boxcars.

▷ O&K Feldbahn, 1903

Wheel arrangement	0-8-0T
Cylinders	2
Boiler pressure	approx. 180 psi (12.65 kg/sq cm)
Drive wheel diameter	approx. 22¾ in (580 mm)
Top speed	approx. 15 mph (24 km/h)

Introduced in 1903, around 2,500 of these 1-ft 11¾-in- (0.60-m-) gauge "Brigadelok" locomotives were built by several German companies, and widely used on the military light railroads constructed to supply forward positions of the German army. The locomotive shown here is No.7999, an Orenstein & Koppel engine built in 1915 with Klein-Linder articulation of the front and rear axles.

△ Baldwin Switcher, 1917

Wheel arrangement	0-6-0T
Cylinders	2
Boiler pressure	190 psi (13.4 kg/sq cm)
Drive wheel diameter	48 in (1,220 mm)
Top speed	approx. 30 mph (48 km/h)

Built in the US by the Baldwin Locomotive Works, the 651–700 Series of Railway Operating Division switching (or shunting) locomotives was introduced in 1917 for use by the British Military Railways in France. After the war, they became Class 58 of the Belgian National Railways.

△ GCR Class 8K, 1911

Wheel arrangement	2-8-0
Cylinders	2
Boiler pressure	180 psi (12.65 kg/sq cm)
Drive wheel diameter	56 in (1,420 mm)
Top speed	approx. 45 mph (72 km/h)

The Great Central Railway's Class 8K freight locomotive introduced in 1911 was chosen as the standard British Railway Operating Division 2-8-0 locomotive during WWI. A total of 521 were built, with many seeing service hauling troop and freight trains in France. During WWII many of these locomotives were sent on active service to the Middle East.

△ **Baldwin ALCO narrow-gauge, 1916**

Wheel arrangement	4-6-0PT
Cylinders	2
Boiler pressure	178 psi (12.51 kg/sq cm)
Drive wheel diameter	23¼ in (590 mm)
Top speed	approx. 18 mph (29 km/h)

Based on a French design, these 1-ft 11¾-in (0.60-m) gauge pannier tank locomotives were supplied by the Baldwin Locomotive Works and the American Locomotive Co. (ALCO) in the US to the British War Office, for use on front-line military railroads in northern France and the Middle East during WWI.

TECHNOLOGY

Armored Engines

The British pioneered the use of small, armored, narrow-gauge gasoline locomotives to operate on the temporary railroads that served the front line during World War I. Unlike steam locomotives, which could easily be spotted by the enemy, these locomotives could haul ammunition trains to forward positions during daylight hours without being detected.

Simplex locomotive Built for the British War Office by Motor Rail Ltd. in 1917, this 1-ft 11¾-in (0.60-m) four-wheel, engine hauled 17-ton (15-metric-ton) ammunition trains at 5 mph (8 km/h) to the trenches in northern France.

△ **Pershing Nord, 1917**

Wheel arrangement	2-8-0
Cylinders	2
Boiler pressure	189 psi (13.28 kg/sq cm)
Drive wheel diameter	56 in (1,420 mm)
Top speed	56 mph (90 km/h)

The North British Locomotive Co. in Glasgow supplied 113 Consolidation Pershings for the Compagnie des Chemins de fer du Nord in France. While the railroad was happy to run these large locomotives at up to 56 mph (90 km/h), other French railroads preferred lower operating speeds.

△ **Baldwin "Spider," 1917**

Wheel arrangement	4-6-0
Cylinders	2
Boiler pressure	190 psi (13.4 kg/sq cm)
Drive wheel diameter	61¾ in (1,570 mm)
Top speed	approx. 65 mph (105 km/h)

Nicknamed "Spiders" by British soldiers, 70 of these mixed-traffic locomotives were built with bar frames by the US Baldwin Locomotive Works between 1917 and 1918 for service on the Western Front during WWI. Later they became Class 40 of the Belgian National Railways.

War Machines

Rail-mounted artillery featured in conflicts from the American Civil War to World War II. It caused great destruction, most notably during World War I, when Germany's *Pariskanonen* ("Paris Guns") bombarded the French capital with 230-lb (106-kg) shells from a distance of 75 miles (120 km). Rail-mounted gun turrets provided fast, mobile, fire power, and being moveable, they could also be hidden from enemy attacks. From 1862 to 1945, before the introduction of air attacks, they were perhaps the most destructive long-range weapons.

ON THE FRONT LINE

Austria-Hungary, Britain, France, Germany, Russia, and the US all deployed rail-mounted artillery. While France was the first nation to equip its army with rail-mounted howitzers by adapting naval guns for use on train cars, Germany developed howitzers that set records for size, range, and destructive power. Companies such as Krupp and Skoda built *Dicke Bertha* (Big Bertha) and *Schlanke Emma* (Skinny Emma) howitzers, respectively, both of which inflicted great damage on French and Belgian defenses. However, Krupp's *Schwerer Gustav* (Heavy Gustav)—the biggest land weapon ever—was a failure. Though capable of firing 4-ton (4,000-kg) shells as far as 29 miles (46 km), it demanded a crew of no fewer than 1,420 people and two parallel tracks, which made the gun so impractical that it only saw action once.

French railroad guns were engaged in the Somme offensive during World War I. The weapon pictured required a crew of 15 men.

Fast and Powerful

The introduction of longer and heavier express passenger trains in Europe and the US during the 1920s and 1930s led to the building of more powerful and faster types of locomotives to standard designs. In Britain, Sir Nigel Gresley led the way with his three-cylinder A1 and A3 Pacific 4-6-2s, of which *Flying Scotsman* is justifiably world famous. Other British locomotive engineers, such as the Great Western Railway's Charles Collett and the London, Midland & Scottish Railway's Henry Fowler, preferred a 4-6-0 wheel arrangement. In the US, Germany, and France, the Pacific became the favored express passenger locomotive type.

△ PRR Class K4s, 1914

Wheel arrangement 4-6-2	
Cylinders 2	
Boiler pressure 205 psi (14.4 kg/sq cm)	
Drive wheel diameter 80 in (2,030 mm)	
Top speed approx. 70 mph (113 km/h)	

The Class K4s Pacific locomotives, of which 425 were built in the US between 1914 and 1928, were the Pennsylvania Railroad's premier express steam locomotive. They were often used in double or triple headers to haul heavy trains.

◁ NZR Class Ab, 1915

Wheel arrangement 4-6-2	
Cylinders 2	
Boiler pressure 180 psi (12.65 kg/sq cm)	
Drive wheel diameter 54 in (1,372 mm)	
Top speed approx. 60 mph (96 km/h)	

One of a class of 141 locomotives, New Zealand Railways Class Ab Pacific locomotive No. 608 is named *Passchendaele* in memory of NZR staff killed in WWI. Ab engines were replaced by diesels in the 1960s but five have been preserved.

◁ SR Class Ps-4, 1923

Wheel arrangement 4-6-2	
Cylinders 2	
Boiler pressure 200 psi (14.06 kg/sq cm)	
Drive wheel diameter 73 in (1,854 mm)	
Top speed approx. 80 mph (129 km/h)	

Finished in a striking green livery, the 64 Class Ps-4 Pacific-type express passenger locomotives were built for the Southern Railway of the US by the American Locomotive Company (ALCO) and the Baldwin Locomotive Works between 1923 and 1928. Designed to haul the railroad's heavy expresses, they had been replaced by diesels by the early 1950s. No. 1401 is on display in the Smithsonian Institution in Washington, DC.

△ PRR Class G5s, 1924

Wheel arrangement 4-6-0	
Cylinders 2	
Boiler pressure 205 psi (14.4 kg/sq cm)	
Drive wheel diameter 68 in (1,730 mm)	
Top speed approx. 70 mph (113 km/h)	

This engine was designed by William Kiesel to work commuter trains on the Pennsylvania Railroad. The Class G5s was one of the largest and most powerful 4-6-0s in the world. No. 5741 is on display in the Railroad Museum of Pennsylvania.

▷ LMS Royal Scot Class, 1927

Wheel arrangement 4-6-0	
Cylinders 3	
Boiler pressure 250 psi (17.57 kg/sq cm)	
Drive wheel diameter 81 in (2,057 mm)	
Top speed approx. 80 mph (129 km/h)	

Designed by Sir Henry Fowler, 70 Royal Scot Class locomotives were built to haul long-distance express trains on the London, Midland & Scottish Railway. They were later rebuilt by William Stanier with Type 2A tapered boilers, and remained in service until the early 1960s.

BUILDING THE 'ROYAL SCOT' ENGINE

◁ DR Class 01, 1926
Wheel arrangement 4-6-2
Cylinders 2
Boiler pressure 232 psi (16.3 kg/sq cm)
Drive wheel diameter 78³/₄in (2,000 mm)
Top speed approx. 81 mph (130 km/h)

A total of 241 (including 10 rebuilt Class 02s) of these standardized Class 01 express locomotives were built for the Deutsche Reichsbahn between 1926 and 1938. Some engines remained in service in East Germany until the early 1980s.

△ LNER Class A3, 1928
Wheel arrangement 4-6-2
Cylinders 3
Boiler pressure 220 psi (15.46 kg/sq cm)
Drive wheel diameter 80 in (2,030 mm)
Top speed 108 mph (174 km/h)

Britain's Sir Nigel Gresley designed the A3 for the London & North Eastern Railway. These locomotives hauled express trains between London's King's Cross and Scotland. No. 4472 *Flying Scotsman* is the only example preserved.

△ GWR Castle Class, 1936
Wheel arrangement 4-6-0
Cylinders 4
Boiler pressure 225 psi (15.82 kg/sq cm)
Drive wheel diameter 80¹/₂in (2,045 mm)
Top speed 100 mph (161 km/h)

These express locomotives were designed by Charles Collett for the Great Western Railway. Its Swindon Works built 171 Castle Class engines between 1923 and 1950. Shown here is No. 5051. They had all been retired by 1965, but eight have now been preserved. No. 5051 *Drysllyn Castle* is at Didcot Railway Centre.

△ GWR King Class, 1930
Wheel arrangement 4-6-0
Cylinders 4
Boiler pressure 250 psi (17.57 kg/sq cm)
Drive wheel diameter 78 in (1,980 mm)
Top speed approx. 90 mph (145 km/h)

The King Class was designed by Charles Collett for the Great Western Railway. Thirty of these express locomotives were built at Swindon Works in England between 1927 and 1936. They were replaced by diesels in the early 1960s; three, including this one, No. 6023 *King Edward II*, have been preserved.

▷ Nord Pacific, 1936
Wheel arrangement 4-6-2
Cylinders 4 (compound)
Boiler pressure 240 psi (16.87 kg/sq cm)
Drive wheel diameter 75¹/₂in (1,918 mm)
Top speed approx. 81 mph (130 km/h)

French engineer André Chapelon designed these powerful locomotives for the Compagnie du Nord. They hauled express trains such as the *Flèche d'Or* in northern France. Shown here is No. 3.1192, which is exhibited at the Cité du Train, Mulhouse, France.

King Edward II

Built at Swindon Works in June 1930, the King Class locomotive No. 6023 *King Edward II* was in a class of engines considered to be the most powerful machines on any British railroad. The first of the class, No. 6000, built in 1927, was named after the reigning monarch, King George V; later engines carried names of earlier kings in reverse order of ascendance. *King Edward II* served for 32 years, first with the Great Western Railway, then British Railways.

DESIGNED BY CHARLES B. COLLETT, the King Class was a natural progression from his four-cylinder 4-6-0 Castle Class engines, which had enjoyed great success. Commentators at the time even wondered whether this was a new design or simply a "super" Castle. The King Class locomotives were able to handle the heaviest trains operated by the GWR, but their heavy axle weight restricted them to the London–Plymouth and London–Wolverhampton (via Bicester) routes. Because of this limited route availability, relatively few were built.

After being withdrawn from service in June 1962, No. 6023 *King Edward II* was sold to locomotive scrap merchants Woodham Brothers of Barry, South Wales, and remained there until its rescue in December 1984. By this time the engine was a rotting hulk and its rear drive wheel set had been sliced through by a cutting torch following a switching mishap. This iconic locomotive has now been fully restored, and it returned to steam at Didcot Railway Centre in 2011.

FRONT VIEW

REAR VIEW

British Railways logo
The original lion emblem, known as the "Cycling Lion," was used on locomotives between 1950 and 1956.

SPECIFICATIONS			
Class	King	**In-service period**	1930-62 (*King Edward II*)
Wheel arrangement	4-6-0	**Cylinders**	4
Origin	UK	**Boiler pressure**	250 psi (17.5 kg/sq cm)
Designer/builder	C.B. Collett/Swindon Works	**Drive wheel diameter**	78 in (1,980 mm)
Number produced	30 King Class	**Top speed**	approx. 110 mph (177 km/h)

Collett tender No. 2460 is one of 24 different tenders to be paired with this locomotive

Double red disks on cab door indicate the engine's GWR power and weight classification

Steel name plate with brass letters

Modified outside steam pipe replaced original pattern

Copper-topped stack is typical of GWR stack design

Lamp bracket on bumper beam

Long service history
King Edward II is one of only three surviving members of its class and performed over 1.500 million miles (2.414 million km) of service. The "PDN" mark on the buffer framing is the GWR code for the Old Oak Common depot where it was first based.

6023

PDN

EXTERIOR

The King Class engines were originally turned out in the GWR's traditional green Swindon livery, with their distinctive, copper-topped stacks. However, in 1948 two King Class locomotives were turned out in an experimental dark blue livery with red, cream, and gray lining. In 1950 a standard Caledonian blue livery with black and white lining was introduced. Over time, British Railways changed the livery back to green. *King Edward II* has been restored in the British Railways 1950s blue livery.

1. Nameplate in brass letters on steel 2. Number plate on cab side 3. Interior of smokebox 4. Stack with polished copper cap 5. Axle and leaf spring suspension on front set of truck wheels 6. Retaining valve for vacuum brake changeover, and copper pipes for lubricator 7. Crosshead of inside cylinder, seen through inspection hole 8. Copper pipes for directing steam from cylinder cocks 9. Crosshead and slidebars 10. Big-end bearing of connecting rod 11. Jacket sheets on side of outer firebox 12. Vacuum brake ejector 13. Builder's plate on rear of tender tank 14. Speedometer drive 15. Low-level tender filler (a modern addition 16. Buffer at rear of tender 17. Front of tender viewed from cab

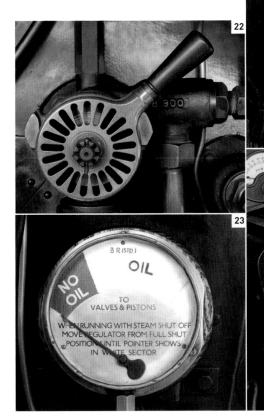

CAB INTERIOR

The King Class footplate layout followed the Swindon Works' standard design, which was practical and reasonably spacious. Early locomotives were generally of the right-hand drive configuration, with the fireman's seat being on the left or near side. When double-track railroads first came into being, the wayside signals were placed on the near side, so many rail companies changed their footplate designs to left-hand drive. However, the GWR continued to configure their locomotives for right-hand driving. Unlike other designers, Collett did not include padded seating for the footplate crew, preferring instead a simple, hinged wooden seat.

18. Cab controls on backhead of firebox
19. Water-level gauge **20.** Firehole door
21. Interior of firebox **22.** Vacuum brake control
23. Mechanical lubricator gauge **24.** Screw reverser (clockwise forward, counterclockwise backward) **25.** Automatic Train Control (ATC) audible signaling system **26.** Wooden seat on fireman's side of cab

Great Journeys
Orient Express

Made famous in literature and film, the *Orient Express* was the brainchild of the Belgian Georges Nagelmackers, founder of the Compagnie Internationale des Wagons-Lits, a company that specialized in operating luxury train services on European railroads.

FOLLOWING A SUCCESSFUL TEST JOURNEY between Paris and Vienna in 1882, the first regular *Express d'Orient* left Gare de l'Est, Paris, behind an outside cylinder Est 2-4-0 locomotive on October 4, 1883. It traveled eastward to Strasbourg and then to Munich before crossing into Austria and calling at Salzburg and Vienna. From there the train continued on to Budapest, Bucharest, and Giurgiu on the banks of the Danube River in Romania. Passengers were then ferried across the river to Rustchuk in Bulgaria, where they boarded older rolling stock of the Austrian Eastern Railway to Varna on the Black Sea coast. From Varna, passengers then made an 18-hour sea voyage to Constantinople. Between Paris and Giurgiu the train consisted of five new sleeping cars, a dining car, and two baggage cars, all built to a high standard in teak, with locomotives changed many times *en route*. The journey took four days in total, so passengers had plenty of time to enjoy the high standard of cuisine in the restaurant car on the first leg of the journey.

From 1889 the train began running directly between Paris and Constantinople, following the opening of new railroads through the Balkans in Serbia, Bulgaria, and European Turkey. The train was renamed the *Orient Express* in 1891.

Services ended with the onset of World War I but recommenced after the war and the train once again became popular. The Simplon Tunnel had opened under the Alps in 1906, and in 1919 a new *Simplon-Orient-Express* took a route between

The early *Orient Express*
This steam locomotive, dating back to 1896, traveled part of the original *Orient Express* route from Paris to Constantinople.

VINTAGE POSTER FROM 1920S

Paris and Constantinople via Milan, Venice, Trieste, and Belgrade. By the 1930s, three separate trains operated: the *Orient Express* on the original 1889 route; the *Simplon-Orient-Express* via the Simplon Tunnel; and the *Arlburg-Orient-Express* via Zurich, Innsbruck, and Budapest with through cars for Athens. Sleeper cars started running from Calais, providing the first transcontinental journey across Europe.

Following suspension during World War II, the *Orient Express* resumed service in 1952, but both it and the *Arlberg-Orient-Express* had ceased to run by 1962. The *Simplon-Orient-Express* was replaced that year by the *Direct-Orient-Express*, which was withdrawn in 1977. Some of the cars were bought in 1982 by a private company, which now runs *Venice-Simplon Orient Express* services to several destinations in Europe.

Splendor on the Orient Express
The saloon car aboard the Orient Express around 1896 was designed for luxury, featuring detailed wood paneling and an inlaid ceiling.

UNITED KINGDOM

NORTH SEA

London

ENGLISH CHANNEL

Calais

Black Forest ☐1
The *Orient Express* passed through the Black Forest mountain range in Baden in southwestern Germany.

Paris

Strasbourg

Zurich

SWITZERLAND

FRANCE

Milan

Monaco

Last journey The last run of the *Direct-Orient-Express* terminated off-route in Monaco in 1977. The 1920s cars were purchased at auction here and later restored, primarily for use on the London to Venice route.

0		150		300 miles
0	150	300	450 km	

N

KEY FACTS

DATES
1883 First regular *Express d'Orient* leaves Gare de l'Est, Paris for Giurgiu in Romania
1889 First through service Paris-Constantinople
1891 Train renamed *Orient Express*
1977 Regular Paris-Istanbul journeys cease

TRAIN
Locomotive In France the first *Orient Express* was hauled by a Chemins de Fer de l'Est outside cylinder 2-4-0. Many different locomotives were used
Cars (1883) 5 truck sleeping cars with accommodations for 20 passengers and 2 washrooms; 1 truck restaurant car; 1 baggage car; 1 mail car

JOURNEY
1883 Paris to Constantinople (original journey)
Train from Paris to Giurgiu; passengers ferried from Giurgiu across Danube to Rustchuk, Bulgaria followed by a 7-hour train to Varna; ship to Constantinople (Istanbul); approx. 1,500 miles (2,410 km), 4 days
1889 Paris to Constantinople
Train diverted at Budapest to Belgrade and Nis, Serbia, through Dragoman Pass in Passara Mountains to Bulgaria, Pazarzhik to Plovdiv, then Constantinople; approx. 1,400 miles (2,250 km), 67 hours 35 minutes
Paris to Istanbul (current, not shown on map)
Runs annually; approx. 1,400 miles (2,250 km) 6 days, 5 nights; spends the night in Budapest and Bucharest
London to Venice (current VSOE route, not on map)
Route via Paris/Innsbruck/Verona; 1,065 miles (1,715 km) 2 days, 1 night

RAILROAD
Gauge Standard 4 ft 8 ½ in (1,435 mm)
Tunnels Longest (*Simplon-Orient-Express* route) is Simplon Tunnel, Alps 65,039 ft (19,824 m)
Highest point Simplon Tunnel, Alps 2,313 ft (705 m)

KEY
● Start/Finish
● Main stations
Original route
1889 route
Change of train
Sea voyage

A ROMANTIC ADVENTURE

The original route of the *Orient Express* is now retraced annually, with all the luxury of the earliest trips. The antique train passes through seven countries, with numerous stops along the way. Plush private cabins, personal stewards, and gourmet meals can all be expected on board.

2 Through the Alps The train passed through the scenic ski resort of St. Anton in the Tyrolean Alps in Austria along the route.

3 Budapest Until unification in 1873, Budapest, the capital of Hungary, was two separate cities: Buda was on the west bank of the Danube River while Pest was on the east bank.

5 Varna The ancient port city of Varna dates back five millennia and its necropolis is the site of the oldest find of gold metallurgy.

Giurgiu Passengers were ferried across the Danube from Giurgiu to board another train at Rustchuck.

Iron Gates 4 This gorge on the Danube forms the boundary between Serbia and Romania.

Sirkeci Station 6 The Orient Express Restaurant is located at the historic Sirkeci train station at Istanbul, the original route's terminus.

Mixed-traffic Movers

By the 1930s the standardization of machine parts by European and US locomotive builders had reduced construction and maintenance costs significantly. Powerful engines designed to haul express freight and passenger trains were soon coming off the production lines in great numbers. In Great Britain, both Charles Collett of the Great Western Railway (GWR) and William Stanier of the London, Midland & Scottish Railway (LMS) made standardization a common theme when designing their new 4-6-0 locomotives, while in Germany the Class 41 2-8-2s built for the Deutsche Reichsbahn incorporated parts simultaneously developed for three other classes.

△ LMS Class 5MT *Black Five*, 1934

Wheel arrangement	4-6-0
Cylinders	2
Boiler pressure	225 psi (15.82 kg/sq cm)
Drive wheel diameter	72 in (1,830 mm)
Top speed	approx. 80 mph (129 km/h)

Designed by William Stanier for the London, Midland & Scottish Railway, many of these powerful mixed-traffic locomotives saw service in Britain until the end of steam in 1968. A total of 842 were built.

◁ SR S15 Class, 1927

Wheel arrangement	4-6-0
Cylinders	2
Boiler pressure	175-200 psi (12.30-14 kg/sq cm)
Drive wheel diameter	67 in (1,700 mm)
Top speed	approx. 65 mph (105 km/h)

These powerful British locomotives were a modified version of an earlier Robert Urie design, introduced by Richard Maunsell. They were built by the Southern Railway at its Eastleigh Works in Southern England.

◁ NZR Class K, 1932

Wheel arrangement	4-8-4
Cylinders	2
Boiler pressure	200 psi (14.06 kg/sq cm)
Drive wheel diameter	54 in (1,372 mm)
Top speed	approx. 65 mph (105 km/h)

Built to haul heavy freight and passenger trains on New Zealand's mountainous North Island, 30 of the Class Ks were built at Hutt Workshops for New Zealand Railways between 1932 and 1936. They were gradually withdrawn from service between 1964 and 1967.

GREAT WESTERN

5900

△ **LNER Class V2, 1936**

Wheel arrangement 2-6-2

Cylinders 3

Boiler pressure 220 psi (15.46 kg/sq cm)

Drive wheel diameter 74 in (1,880 mm)

Top speed approx. 100 mph (161 km/h)

These engines were designed by Sir Nigel Gresley for the London & North Eastern Railway and hauled both express passenger and express freight trains. No. 4771 *Green Arrow* is the only preserved example.

Express Freight

Transporting perishable goods such as milk, fish, and meat by rail called for specialized freight cars. In Britain, milk was first conveyed in milk churns loaded into ventilated wagons at country stations, but from the 1930s it was carried in six-wheel milk tank wagons loaded at a creamery. The wagons were marshaled into trains and hauled by powerful express steam locomotives to depots in and around London. The last milk trains to operate in Britain ceased in 1981.

London's dairy supplier With a capacity of 3,600 gallons (13,638 liters), the Express Dairy six-wheel milk tank wagon weighed as much as a loaded passenger coach when full. This wagon was built by Southern Railway in 1931 and rebuilt in 1937.

△ **DR Class 41, 1937**

Wheel arrangement 2-8-2

Cylinders 2

Boiler pressure 290 psi/228 psi (20.39 kg/sq cm/16 kg/sq cm)

Drive wheel diameter 63 in (1,600 mm)

Top speed approx. 56 mph (90 km/h)

Built with parts that were designed for several different locomotive types, these powerful, fast freight engines were constructed for the Deutsche Reichsbahn between 1937 and 1941.

◁ **GWR Manor Class, 1938**

Wheel arrangement 4-6-0

Cylinders 2

Boiler pressure 225 psi (15.82 kg/sq cm)

Drive wheel diameter 68 in (1,730 mm)

Top speed approx. 65 mph (105 km/h)

With their light axle loading, these Great Western Railway mixed-traffic locomotives could operate on secondary and branch lines as well as main lines in England and Wales. This engine is No. 7808 *Cookham Manor*.

◁ **GWR Hall Class, 1928**

Wheel arrangement 4-6-0

Cylinders 2

Boiler pressure 225 psi (15.82 kg/sq cm)

Drive wheel diameter 72 in (1,830 mm)

Top speed approx. 70 mph (113 km/h)

A total of 259 of these versatile engines, designed by Charles Collett, were built at the Great Western Railway's Swindon Works between 1928 and 1943. This is No. 5900 *Hinderton Hall*.

Versatile Engines

While the development of more powerful and faster express steam locomotives gathered pace during the 1920s and 1930s, there was also the parallel development of smaller engines designed for switching (or shunting) at freight yards, railroad workshops, and stations, or to carry out passenger and freight duties on rural branch lines. Many of these versatile locomotives remained in active service until the end of the steam era, while some have since been restored to service on heritage railroads.

▷ P&R Switcher No. 1251, 1918

Wheel arrangement 0-6-0T

Cylinders 2

Boiler pressure 150 psi (10.53 kg/sq cm)

Drive wheel diameter 50 in (1,270 mm)

Top speed approx. 25 mph (40 km/h)

Rebuilt in 1918 from a Class 1-2a Consolidation locomotive, No. 1251 spent its life as a switcher at the Philadelphia & Reading Railroad Shops in Reading, Pennsylvania. It was retired in 1964 as the last steam engine on a US Class 1 railroad, and is now on display at the Railroad Museum of Pennsylvania.

△ LMS Class 3F "Jinty," 1924

Wheel arrangement 0-6-0T

Cylinders 2 (inside)

Boiler pressure 160 psi (11.25 kg/sq cm)

Drive wheel diameter 55 in (1,400 mm)

Top speed approx. 40 mph (64 km/h)

These tank locomotives, nicknamed "Jintys," were designed by Henry Fowler for the London, Midland & Scottish Railway. Widely used for switching and local freight work in the Midlands and northwest England, 422 were built, with the last examples remaining in service until 1967.

△ GWR 5600 Class, 1924

Wheel arrangement 0-6-2T

Cylinders 2 (inside)

Boiler pressure 200 psi (14.06 kg/sq cm)

Drive wheel diameter 55½ in (1,410 mm)

Top speed approx. 45 mph (72 km/h)

Designed for the Great Western Railway by Charles Collett, 150 of these powerful tank engines were built at the company's Swindon Works and 50 by Armstrong Whitworth in Newcastle-upon-Tyne. They mainly saw service in the South Wales valleys hauling coal trains, but were also used on local passenger services.

▽ L&B *Lew*, 1925

Wheel arrangement 2-6-2T

Cylinders 2

Boiler pressure 160 psi (11.25 kg/sq cm)

Drive wheel diameter 33 in (840 mm)

Top speed approx. 25 mph (40 km/h)

Completed at the Ffestiniog Railway's Boston Lodge Works in 2010, *Lyd* (shown) is a replica of *Lew*, which was built by Manning Wardle in 1925 for the Southern Railway's 1-ft 11¾-in- (0.60-m-) gauge Lynton to Barnstaple line. The line, closed in 1935, is now in the process of being reopened by enthusiasts.

TECHNOLOGY

Battery Power

Battery locomotives are powered by huge onboard batteries that are recharged in between duties. These engines were once used on railroads serving industrial complexes, such as explosives and chemical factories, mines, or anywhere else where normal steam or diesel locomotives could present hazards, such as fire risk, explosion, or fumes. In England, the London Underground uses battery-electric locomotives when the normal electric power is turned off during periods of nighttime maintenance.

English Electric EE788 0-4-0 Battery Locomotive This four-wheel, 70-hp (52-kW), battery-electric locomotive was built by English Electric at their Preston factory in England in 1930 and worked for many years at their Stafford Works. It is currently on display at the Ribble Steam Railway Museum in Preston.

△ GWR 4575 Class Prairie Tank, 1927

Wheel arrangement 2-6-2T

Cylinders 2

Boiler pressure 200 psi (14.06 kg/sq cm)

Drive wheel diameter 55½ in (1,410 mm)

Top speed approx. 50 mph (80 km/h)

Designed by Charles Collett, the 4575 Class of Prairie tank was built at the Great Western Railway's Swindon Works between 1927 and 1929. Of the 100 built, many saw service on branch line passenger and freight duties in England's West Country. No. 5572 shown here was one of six equipped for push-pull operations. It is preserved at Didcot Railway Centre.

△ GWR 5700 Class Pannier Tank, 1929

Wheel arrangement 0-6-0PT

Cylinders 2 (inside)

Boiler pressure 200 psi (14.06 kg/sq cm)

Drive wheel diameter 55½ in (1,410 mm)

Top speed approx. 40 mph (64 km/h)

One of the most numerous classes of British steam engine, 863 of these Pannier Tanks were built for the Great Western Railway and British Railways between 1929 and 1950. They were usually seen at work on switching duties or hauling passenger and freight trains on branch lines. Of the 16 preserved, No. 3738, seen here, is on display at Didcot Railway Centre.

△ DR Class 99.73-76, 1928

Wheel arrangement 2-10-2T

Cylinders 2

Boiler pressure 200 psi (14.06 kg/sq cm)

Drive wheel diameter 31½ in (800 mm)

Top speed approx. 19 mph (31 km/h)

The Deutsche Reichsbahn had these tank engines built as a new standard design for 2-ft 5½-in- (0.7-m-) gauge lines in Saxony, eastern Germany. A number of these and a modified version introduced in 1950s are still in service today.

△ EIR Class XT/1, 1935

Wheel arrangement 0-4-2T

Cylinders 2

Boiler pressure 160 psi (11.25 kg/sq cm)

Drive wheel diameter 57 in (1,448 mm)

Top speed approx. 40 mph (64 km/h)

Built by Friedrich Krupp AG of Berlin, Germany, for the 5-ft 6-in- (1.67-m-) gauge East Indian Railway, these locomotives were first introduced in 1929 and were used for light passenger work. No. 36863 (shown) was built in 1935 and is on static display at the National Rail Museum, New Delhi.

Freight Shifters

As train speeds rose, they increasingly carried a variety of goods, including perishable food items. Freight locomotives evolved accordingly. Mainland Europe and North America discarded the six-wheeler for front-rank duties, but the UK continued to build them. The 2-8-0, and variants on the eight-coupled wheelbase, became the main types. Canada, China, Germany, and the USSR built 10-coupled designs, but, especially in the US, the loads and terrain demanded nothing short of the giants.

◁ PRR Class A5s, 1917

Wheel arrangement	0-4-0
Cylinders	2
Boiler pressure	185 psi (13 kg/sq cm)
Drive wheel diameter	50 in (1,270 mm)
Top speed	approx. 25 mph (40 km/h)

The Pennsylvania Railroad served many industrial sites around Baltimore, Philadelphia, and New York, where a short-wheelbase switcher, or shunter, was essential to negotiate the tight clearances. One of the most powerful 0-4-0s ever, 47 of the Class A5s were built at the railroad's workshops in Altoona, PA, up to 1924.

△ XE Class, 1928/30

Wheel arrangement	2-8-2
Cylinders	2
Boiler pressure	210 psi (14.76 kg/sq cm)
Drive wheel diameter	61½ in (1,562 mm)
Top speed	approx. 30 mph (48 km/h)

Aside from articulated types, the XE (X Eagle) Class of British-built Mikados (2-8-2s) were the largest steam locomotives on the subcontinent. A total of 93 of these broad-gauge (5-ft 6-in/1.67-m) designs were built, of which 35 were based in Pakistan after partition. No. 3634 *Angadh* is shown here.

▷ CP T1-C Class Selkirk, 1929

Wheel arrangement	2-10-4
Cylinders	2
Boiler pressure	285 psi (20.03 kg/sq cm)
Drive wheel diameter	63 in (1,600 mm)
Top speed	approx. 65 mph (105 km/h)

This semi-streamlined class of engines was built by Canadian Pacific Railway to master the Selkirk Mountains. Thirty of these oil-burners were built up to 1949. They were the largest and most powerful non-articulated locomotives in the British Commonwealth, and they hauled trains 262 miles (422 km) over the mountains from Calgary, Alberta, to Revelstoke, British Columbia.

▷ DR Class 44, 1930

Wheel arrangement	2-10-0
Cylinders	3
Boiler pressure	228 psi (16.02 kg/sq cm)
Drive wheel diameter	55 in (1,397 mm)
Top speed	approx. 50 mph (80 km/h)

The Deutsche Reichsbahn acquired the first 10 in 1926, but delayed further orders until 1937, after which no fewer than 1,979 were built up to 1949. Unusually for a freight design they had three cylinders, letting them haul trains of up to 1,323 tons (1,200 metric tons).

Goods Wagons

By the 20th century, railroads hauled loads ranging from salt to sugar, gasoline to milk, and cattle to coal. Wagons evolved to fill specific roles: hoppers transported coal, ores, and stone; tankers carried liquids and gases; and refrigerated cars carried perishable goods. Whatever the load, before the introduction of continuous braking, every train had a brake van. From here the conductor kept watch over the train, using his brake to keep control of the loose-coupled wagons on down gradients and when stopping.

△ GWR "Toad" brake van, 1924

Type	Brake van
Weight	22.4 tons (20.32 metric tons)
Construction	wooden body on four-wheel steel chassis
Railroad	Great Western Railway

At a time when most British freight trains lacked any form of through braking, the role of the conductor (or brakeman) was critical. From 1894 the Great Western Railway's conductors manned "Toads," the name deriving from the electric telegraph code for brake vans.

◁ **UP Challenger CSA-1 Class/ CSA-2 Class, 1936**

Wheel arrangement	4-6-6-4
Cylinders	4
Boiler pressure	280 psi (19.68 kg/sq cm)
Drive wheel diameter	69 in (1,753 mm)
Top speed	approx. 70 mph (113 km/h)

Union Pacific Railroad's Challenger proved that a simple articulated engine could haul huge loads at high speed. Each set of drive wheels was powered by two cylinders, with four trailing wheels to support the huge firebox. The American Locomotive Company (ALCO) built 105 from 1936 to 1944. Two have been preserved, No. 3977 and No. 3985.

▷ **SAR Class 15F, 1938**

Wheel arrangement	4-8-2
Cylinders	2
Boiler pressure	210 psi (14.76 kg/sq cm)
Drive wheel diameter	60 in (1,524 mm)
Top speed	approx. 60 mph (96 km/h)

Most numerous of South African Railway's classes, the 15F was used predominantly in the Orange Free State and Western Transvaal. Construction spanned WWII; 205 were built by UK companies and a further 50 by German ones. Several have survived. The 1945-built No. 3007 is in the city of its birth at Glasgow's Riverside Museum.

◁ **GWR 2884 Class, 1938**

Wheel arrangement	2-8-0
Cylinders	2
Boiler pressure	225 psi (15.81 kg/sq cm)
Drive wheel diameter	55½ in (1,410 mm)
Top speed	approx. 45 mph (72 km/h)

The Great Western Railway's 2800 Class of 1903—the first British 2-8-0—was a success, persuading the GWR to add to the original total of 83. Modifications, though minor, merited a new designation—the 2884 Class, 81 of which were built from 1938 to 1942. No. 3822 is one of nine preserved.

◁ **FGEX fruit boxcar, 1928**

Type	Express refrigerated boxcar
Weight	27.7 tons (25.13 metric tons)
Construction	wooden body with integral cooling system mounted on steel underframe with two four-wheel trucks
Railroad	Fruit Growers' Express

A leasing company jointly owned by 11 railroads in the eastern and southeastern US, the Fruit Growers' Express built and operated several thousand refrigerated vehicles. Retired in the late 1970s, No. 57708 was preserved by the Cooperstown & Marne Railroad.

△ **ACF three-dome tanker, 1939**

Type	Three-dome truck oil tanker
Weight	20.3 tons (18.37 metric tons)
Construction	steel superstructure mounted on a double truck steel chassis
Railroad	The Shippers' Car Line Corporation

The American Car & Foundry Company remains one of the major rolling stock manufacturers in the US. It built three-dome tanker No. 4556 in 1939 for the Shippers' Car Line Corporation. Riding on two four-wheel trucks, and used for transporting propane and liquid petroleum gas, the tanker has a capacity of 4,550 gallons (17,230 liters).

Herbert Nigel Gresley
1876-1941

Nigel Gresley started his engineering career at the age of 17, when he became an apprentice at Crewe Locomotive Works. After serving his apprenticeship, he broadened his experience in the field by working as a maintainer, designer, and tester, as well as the foreman of an engine barn. In 1905 he began working for the Great Northern Railway, where he designed locomotives, pioneered articulated cars, and eventually rose to the position of Locomotive Superintendent in 1911. After the formation of the London and North Eastern Railway (LNER) in 1923, Gresley was appointed its Chief Mechanical Engineer, a post he held until his death. He was knighted in 1936.

ENGINEER AND INNOVATOR

Gresley initially started work on the design for a Pacific in 1915, but when his first was actually built in 1922, it was a very different machine. By then Gresley had developed a conjugated valve gear that simplified the drive from three-cylinder engines. Gresley went on to design Great Britain's largest and most powerful steam locomotive, the 1925 Garratt 2-8-0+0-8-2, and its largest passenger steam locomotive, the Class P2 2-8-2. In an effort to increase efficiency, he also experimented with a high-pressure water-tube boiler originally developed for ships.

From 1928 Gresley developed the A1 Pacifics into A3 Pacifics, adding higher-pressure boilers to further improve performance. However, the first recorded steam locomotive speed of 100 mph (161 km/h) was made by an A1 Pacific, *Flying Scotsman*, on November 30, 1934. The next year, Gresley introduced the A4 Pacific, with elegant streamlined styling. It was A4 Pacific *Mallard* that set the current steam locomotive speed record in 1938.

Despite his achievements in steam, Gresley remained open to other methods of rail propulsion and in 1936 he began designs for trans-Pennine electrification using 1,500 V DC locomotives. Delayed by World War II, the project was completed in the 1950s.

Record-breaking steam
Mallard was the ultimate expression of Gresley's A4 Pacific: on July 3, 1938, it set a world steam record speed of 126 mph (203 km/h) that has never been beaten. The locomotive survives at the National Railway Museum, York, England.

Streamlined Steam Around Europe

The 1930s was the Golden Age of high-speed, steam-hauled trains in Europe. With national pride at stake, railways competed for the coveted title of the world's fastest train. In Britain the Great Western Railway's *Cheltenham Flyer* was first off the mark in 1932. Hauled by Sir Nigel Gresley's new streamlined, A4 Pacifics, the London & North Eastern Railway's *Silver Jubilee* (1935) and *Coronation* (1937) services set new standards in speed, luxury, and reliability. Steam speed records continued to be broken, first by the German Class 05 in 1936 and then by Gresley's *Mallard* in 1938. World War II ended this high-speed excitement, although *Mallard*'s record has never been broken.

△ LNER Class P2, 1934

Wheel arrangement	2-8-2
Cylinders	3
Boiler pressure	220 psi (15.46 kg/sq cm)
Driving wheel diameter	74 in (1,880 mm)
Top speed	approx. 75 mph (121 km/h)

Sir Nigel Gresley's Class P2 locomotives hauled heavy express passenger trains between London and Aberdeen. Six of the powerful engines were built at the London & North Eastern Railway's Doncaster Works between 1934 and 1936. The class was rebuilt as Class A2/2 Pacifics during WWII.

△ LMS Coronation Class, 1938

Wheel arrangement	4-6-2
Cylinders	4
Boiler pressure	250 psi (17.57 kg/sq cm)
Driving wheel diameter	81 in (2,057 mm)
Top speed	approx. 114 mph (183 km/h)

Designed by William Stanier, a total of 38 of these powerful express locomotives were built at the London, Midland & Scottish Railway's Crewe Works between 1937 and 1948. Ten were built with a streamlined casing that was removed after WWII. No. 6229 *Duchess of Hamilton*, refitted with its streamlined casing, has been preserved.

△ DR Class 05, 1935

Wheel arrangement	4-6-4
Cylinders	3
Boiler pressure	290 psi (20.39 kg/sq cm)
Driving wheel diameter	90½ in (2,299 mm)
Top speed	125 mph (201 km/h)

Three of these streamlined Class 05 express passenger locomotives were built for the Deutsche Reichsbahn in Germany between 1935 and 1937. During 1936 No. 05.002 set a world speed record for steam locomotives of 125 mph (201 km/h) between Berlin and Hamburg. No. 05.001 is preserved in Nürnburg.

▷ LNER Class A4, 1935

Wheel arrangement	4-6-2
Cylinders	3
Boiler pressure	250 psi (17.57 kg/sq cm)
Driving wheel diameter	80 in (2,030 mm)
Top speed	126 mph (203 km/h)

British engineer Sir Nigel Gresley designed the Class A4 streamlined locomotive. Thirty-five of them were built at the London & North Eastern Railway's Doncaster Works between 1935 and 1938. No. 4468 *Mallard* set a still-unbeaten world speed record for steam engines of 126 mph (203 km/h) on the East Coast Main Line in 1938.

△ **SNCB Class 12, 1938**

Wheel arrangement	4-4-2
Cylinders	2 (inside)
Boiler pressure	256 psi (18 kg/sq cm)
Driving wheel diameter	82½ in (2,096 mm)
Top speed	103 mph (166 km/h)

The Class 12 was designed by Raoul Notesse for the Belgian state railways. Six of these Atlantic-type locomotives were built between 1938 and 1939 to haul the Brussels to Ostend boat trains. They were retired in 1962 and No. 12.004 has since been preserved.

TALKING POINT

Traveling Exhibit

The London, Midland & Scottish Railway's streamlined *Coronation Scot* train was shipped across the Atlantic to appear in Baltimore, US. It traveled over 3,000 miles (4,828 km) around the US before being exhibited at the New York World's Fair in 1939. It was unable to return to Britain because of the onset of World War II. The locomotive, No. 6229 *Duchess of Hamilton*, masquerading as No. 6220 *Coronation*, was eventually shipped back to the UK in 1942 but the coaches remained in the US where they were used by the US Army as an officer's mess until after the war, when they too were returned.

Duchess of Hamilton's headlamp One of the headlamps on *Duchess of Hamilton*, this lamp remained in the US and is now on display at the Baltimore & Ohio Railroad Museum in Baltimore.

◁ **DR Class 03.10, 1939**

Wheel arrangement	4-6-2
Cylinders	3
Boiler pressure	290 psi (20.39 kg/sq cm)
Driving wheel diameter	78¾ in (2,000 mm)
Top speed	87 mph (140 km/h)

A total of 60 of these streamlined express passenger locomotives were built for the Deutsche Reichsbahn between 1939 and 1941. After WWII the class was split between East and West Germany and Poland. The German locomotives were rebuilt without their streamline casing, retiring in the late 1970s.

TECHNOLOGY

The Silver Jubilee Service

Named to honour the 25-year reign of King George V, the *Silver Jubilee* high-speed express train was introduced by the London & North Eastern Railway between London King's Cross and Newcastle-upon-Tyne in 1935. Painted in two-tone silver and gray, the articulated train was hauled by one one of Sir Nigel Gresley's Class A4 streamlined Pacific locomotives. The first four were called *Silver Link*, *Quicksilver*, *Silver King*, and *Silver Fox*. The service ceased on the onset of World War II.

Inaugural run LNER Class A4 No. 2509 *Silver Link* departs King's Cross station with the inaugural *Silver Jubilee* express to Newcastle on 30 September 1935.

Mallard

During the 1930s, the desire to lay claim to the fastest locomotive travel times dominated the industry. Top speed records regularly changed hands between the industrialized nations. Then, on July 3, 1938, one of Sir Nigel Gresley's A4 Class Pacific steam engines, LNER No. 4468 *Mallard*, achieved 126 mph (203 km/h), winning Britain the world steam record. The start of World War II ended such record attempts, and *Mallard*'s feat has never been beaten.

THE GRESLEY A4 CLASS streamlined 4-6-2 Pacific locomotive built for the London & North Eastern Railways (LNER) was heralded as an iconic British engine design. The LNER had wanted to speed up their services and their Chief Mechanical Engineer, Nigel Gresley, had observed streamlined trains during a trip to Germany. After discussion in 1935, the LNER board gave Gresley and his design team the go-ahead to develop streamliners especially for the railroad.

The first of the resulting 3-cylinder streamlined A4 class locomotives, No. 2509 *Silver Link,* was completed at Doncaster Works in September 1935; No. 4468 *Mallard* followed in March 1938. Gresley is said to have modeled their impressive streamlined casing on a wedge-shaped Bugatti railcar he had seen in France. Their futuristic look certainly attracted much publicity. Although the A4 class was in steam-engineering terms a development of Gresley's earlier A3 class Pacific types, its strikingly slick casing could not have made it look more different.

SIDE VIEW

FRONT VIEW

SPECIFICATIONS				
Class	A4		In-service period	1938-66 (*Mallard*)
Wheel arrangement	4-6-2 (Pacific)		Cylinders	3
Origin	UK		Boiler pressure	250 psi (17.57 kg/sq cm)
Designer/builder	Sir Nigel Gresley/Doncaster Works		Drive wheel diameter	80 in (2,030 mm)
Number produced	35 A4s		Top speed	126 mph (203 km/h)

Tender had capacity for 8.8 tons of coal and 6,000 gallons (22,730 liters) of water

Corridor tender allowed footplate crews to change on nonstop long runs

State-of-the-art cab gave crew protection in all weather

Smooth casing helped clear smoke away from driver's field of vision

Double stack was first introduced on *Mallard*

Streamlined shape improved aerodynamics

Honoring speed record
A plaque on the locomotive shows the record speed as 126 mph (203 km/h)—more than 2 miles (3.22 km) per minute. It was measured by timekeepers traveling in a dynamometer car. During the record run, the middle big end ran hot, causing bearing metal to melt.

Serving the Silver Jubilee
The A4 class locomotives were built to haul the LNER's *Silver Jubilee* high-speed passenger service from London to Newcastle. On the outside, the engine's streamlined shape both improved speed and reduced fuel consumption, while, on the inside, streamlined ports allowed steam to flow freely. *Mallard* was the first A4 to be equipped with the Kylchap exhaust and blast pipe. The sideskirts were removed during WWII to aid maintenance and were replaced only during the later restoration.

MALLARD

N? 4468

CLASS

EXTERIOR

The streamlined design and smooth casing of *Mallard* not only offered a speed advantage but also helped clear smoke away quickly. The engine had a wedge-shaped front end with a door built into it to allow access to the smokebox for servicing purposes, in particular for clearing out ash; the door's shape earned it the nickname "cod's mouth." *Mallard's* innovative Kylchap exhaust system was located beneath its double stack, which was so successful that it led to the whole class of engines being rebuilt with this type of chimney. The A4's unique sideskirts, or valances, were designed by engineer Oliver Bulleid.

1. Metal nameplate **2.** Engine number and class, handpainted on the nose **3.** Front bumper **4.** Whistle **5.** Aerodynamic stack **6.** Coupling hook at front **7.** Brass builder's plaque **8.** Drive wheel **9.** Detail of outside connecting rod big end **10.** Drive wheel return crank **11.** Axle box and cover **12.** Leaf spring suspension

CAB INTERIOR

The locomotive cab was roofed to give the crew a degree of protection from the elements. They had to work in concert in the small space to get the most out of the engine. From his drop-down seat, the engineer controlled the steam delivered to the cylinders using the regulator valve. The fireman shoveled coal onto the 41¼ sq ft (3.83 sq m) grate of the firebox, and made sure the boiler always contained the right amount of water. Tenders had an 18-in- (45-cm-) wide corridor so that engine crews could change over while the train was moving.

13. Cab controls and backhead of boiler   14. Brake controls
15. Reverser control   16. Vacuum gauge and steam chest gauge
17. Injector control   18. Blower shutoff valve, left, pressure gauge shutoff valve, right   19. Water level gauge   20. Boiler pressure gauge
21. Firebox door   22. Cylinder-cock control   23. Water control lever for injectors   24. Flaman speed recorder   25. Engineer's seat
26. Access door to coal space   27. Plaque attached to rear of tender with instructions for use of the water scoop   28. Tender coal space

The Age of Speed and Style

Symbolized by the futuristic designs of the trains, planes, and automobiles of that period, the decade before World War II could rightly be called "The Age of Speed." Across the world, railroad companies were introducing modern high-speed expresses designed to entice the traveling public on board with their luxurious interiors, slick service, and dependable, fast schedules. Apart from a few diesel-powered streamliners in Germany and the US, these iconic trains were hauled by the latest Art Deco–style steam locomotives, many designed by some of the world's leading industrial designers.

△ VR S Class, 1937

Wheel arrangement	4-6-2
Cylinders	3
Boiler pressure	200 psi (14.06 kg/sq cm)
Drive wheel diameter	73 in (1,854 mm)
Top speed	86 mph (138 km/h)

First introduced in 1928, the four Australian Victoria Railways S Class Pacific-type locomotives were given a streamlined casing in 1937 to haul the new nonstop, Art Deco–style *Spirit of Progress* express between Melbourne and Albury. They had all been scrapped by 1954 after the introduction of diesels.

▷ Japan/China Class SL7, 1935

Wheel arrangement	4-6-2
Cylinders	2
Boiler pressure	220 psi (15.46 kg/sq cm)
Drive wheel diameter	78³/₄ in (2,000 mm)
Top speed	87 mph (140 km/h)

Built by Kawasaki Heavy Industries in Japan and the Shahekou Plant in the Kwantung Leased Territory in China, the 12 Pashina-type locomotives hauled the *Asia Express* during Japanese control of the South Manchuria Railway between 1934 and 1943. Designated Class Shengli 7 after the war, they remained in service in China until the 1970s.

▽ MILW Class A, 1935

Wheel arrangement	4-4-2
Cylinders	2
Boiler pressure	300 psi (21.09 kg/sq cm)
Drive wheel diameter	84 in (2,134 mm)
Top speed	112¹/₂ mph (181 km/h)

Designed to haul *Hiawatha* expresses, four of these high-speed Atlantic-type Class A locomotives were built for the Milwaukee Road (MILW) from 1935 to 1937. Locomotive "A" No. 2 achieved 112¹/₂ mph (181 km/h) between Milwaukee and New Lisbon, WI, in May 1935.

Rail and Road

By the mid-1930s, American Art Deco-style cars and streamlined steam trains were capable of achieving speeds of 120 mph (193 km/h). Industrial designers such as the American Gordon Buehrig, the Franco-American Raymond Loewy, the Englishman John Gurney Nutting, and the Italian-born Frenchman Ettore Bugatti all left their mark on the brief but exciting period of technological progress that ended with the start of World War II.

Speed rivalry Now highly sought after, Jack Juratovic's iconic "Road and Track" prints of 1935 feature a Duesenberg Torpedo Phaeton car racing a streamlined steam train.

△ CN Class U-4-a, 1936

Wheel arrangement	4-8-4
Cylinders	2
Boiler pressure	275 psi (19.33 kg/sq cm)
Drive wheel diameter	77 in (1,956 mm)
Top speed	90 mph (145 km/h)

Five of these streamlined Confederation-type express passenger locomotives were built for Canadian National Railways by the Montreal Locomotive Works in 1936. They remained the premier express locomotives between Toronto and Montreal until replaced by diesels in the 1950s.

△ NSWGR Class C38, 1943

Wheel arrangement	4-6-2
Cylinders	2
Boiler pressure	245 psi (17.22 kg/sq cm)
Drive wheel diameter	69 in (1,750 mm)
Top speed	80 mph (129 km/h)

Designed in 1939 five of the standard-gauge Australian Class C38 express passenger locomotives were actually delivered to the New South Wales Government Railways by Clyde Engineering of Sydney between 1943 and 1945. After hauling expresses, they were retired between 1961 and 1976.

△ PP&L "D" Fireless locomotive, 1939

Wheel arrangement	0-8-0
Cylinders	2
Boiler pressure	130 psi (9.14 kg/sq cm)
Drive wheel diameter	42 in (1,067 mm)
Top speed	20 mph (32 km/h)

Streamlined, but not fast, this Pennsylvania Power & Light Co. fireless switcher was built by Heisler for the Hammermill Paper Co. in Erie. Used in industrial plants where inflammable fuel would be a hazard, fireless locomotives stored steam in their boilers. The largest of this type built, No. 4094-D is on display in the Railroad Museum of Pennsylvania in Strasburg, PA.

Raymond Loewy

Nicknamed "the father of Streamlining," French-born Raymond Loewy (1893–1986) was an American industrial designer known for his wide-ranging work for US industry. In addition to designing world-famous logos for oil companies, such as Shell, and railroads, he also left his mark on Studebaker cars and iconic railroad locomotives such as the Pennsylvania Railroad's Class K4s, T1 and S1 streamlined steam engines. After opening an office in London in 1930, Loewy went on to restyle the Baldwin Locomotive Co.'s early diesel locomotives. Loewy returned to live in his native France in 1980 and died a few years later.

Standing tall Raymond Loewy stands on one of his iconic designs, Pennsylvania Railroad's unique Class S1 6-4-4-6 experimental streamliner locomotive, the US's largest and fastest high-speed locomotive.

Diesel and Electric Streamliners

The 1930s saw the introduction of high-speed diesel and electric trains in Europe and North America. Designed by leading engineers such as Ettore Bugatti and tested in wind tunnels, these streamliners caught the public's imagination, broke world speed records, and ushered in the new age of high-speed rail travel. In Europe, the Germans led the way with their *Flying Hamburger*, the forerunner of today's intercity expresses, and in the US the *Pioneer Zephyr* reached new heights of futuristic modern design. The onset of World War II brought an abrupt end to this exciting progress.

△ DR Class SVT 137
Fliegender Hamburger, 1932

Wheel arrangement	two-car articulated set—front and rear trucks 2′ Bo′ 2′
Transmission	each car electric (1 traction motor)
Engine	each car Maybach 12-cylinder diesel 8,850 cc
Total power output	810 hp (604 kW)
Top speed	99 mph (160 km/h)

With a prototype built in 1932, the Deutsche Reichsbahn train entered service in 1935 between Berlin and Hamburg; it had a buffet and seated 98. The streamlined, two-car diesel-electric *Fliegender* (flying) *Hamburger* was the fastest regular service in the world at an average speed of 77 mph (124 km/h). Inactive during WWII, it worked in France in 1945–49 before returning to operate in Germany until 1983.

△ SBB Class Ae8/14, 1931

Wheel arrangement	(1′A)A1A(A1′) + (1A′)A1A(A1′)
Power supply	15 kV 17 Hz AC, catenary
Power rating	7,394–10,956 hp (5,515–8,173 kW)
Top speed	62 mph (100 km/h)

Three prototype Class Ae8/14 electric locomotives were built for the Swiss Federal Railways' (Schweizerische Bundesbahnen, or SBB) Gotthard line in the 1930s. Each of these powerful double locomotives had eight driving axles and could haul heavy trains unaided over this difficult route. No. 11852 (shown) was for a time the most powerful locomotive in the world.

▷ Bugatti railcar (autorail), 1932/33

Wheel arrangement	each car 2 x 8-wheel trucks, 2 or 4 axles powered
Transmission	mechanical
Engine	each car 2 or 4 Bugatti 12,700 cc
Total power output	4 engines 800 hp (596 kW)
Top speed	122 mph (196 km/h)

Designed by Ettore Bugatti and built in the Bugatti factory in Alsace, France, these gasoline-engined railcars were supplied as single-, double-, or triple-car units. The most comfortable and fastest was the 48-seat, two-car, four-engined "Presidentiel," which set a world rail-speed record of 122 mph (196 km/h) in 1934.

▷ GWR streamlined railcar, 1934

Wheel arrangement	2 x 4-wheel trucks, 1 powered
Transmission	mechanical
Engine	8,850 cc AEC diesel
Total power output	130 hp (97 kW)
Top speed	approx. 63 mph (100 km/h)

First introduced by the Great Western Railway in 1934, these streamlined diesel railcars were nicknamed "Flying Bananas" and remained in service on British Railways until the early 1960s. Production versions, including parcel cars and articulated buffet sets, were equipped with two AEC diesel engines, allowing a top speed of 80 mph (129 km/h).

TECHNOLOGY

German Experiment

The *Schienenzeppelin*, or "rail zeppelin," was an experimental railcar with an aluminum body, which looked like a Zeppelin airship. The front-end design of this prototype bore an uncanny resemblance to the Japanese *Bullet Train* of the 1960s.

Weighing only 20 tons, this 85-ft (26-m) long propeller-driven car was powered by a BMW 12-cylinder gasoline aircraft engine producing a power of 600 hp (447 kW). In June 1931 it set a world land-speed record for rail vehicles using air propulsion when it reached 143 mph (230 km/h) on the Berlin to Hamburg line. The railcar was scrapped in 1939 to provide material for the German war effort in World War II.

Zeppelin train Built by Franz Kruckenberg of Hannover, the *Schienenzeppelin* had only two axles and was designed to carry 40 passengers.

Rear fairing The wind-tunnel-designed fairing had a four-bladed propeller made of ash wood.

◁ **PRR Class GG1, 1934**

Wheel arrangement	2-C+C-2
Power supply	11 kV 25 Hz AC, catenary
Power rating	4,620 hp (3,446 kW)
Top speed	approx. 100 mph (161 km/h)

A total of 139 of these powerful electric locomotives, nicknamed "Blackjacks," were built for the Pennsylvania Railroad between 1934 and 1943. They entered service in 1935, hauling express passenger trains on the newly electrified New York to Washington, DC line. Relegated to freight service in the 1950s, they had all been withdrawn by 1983. No. 4935 is preserved at the Railroad Museum of Pennsylvania.

◁ **CB&Q _Pioneer Zephyr_, 1934**

Wheel arrangement	3 x articulated cars on 4 trucks
Transmission	mechanical
Engine	8-cylinder Winton diesel
Total power output	600 hp (447 kW)
Top speed	112¹/₂ mph (181 km/h)

Built by the Budd Co. for the Chicago, Burlington & Quincy Railroad, the _Pioneer Zephyr_ was a streamlined train of three stainless-steel cars articulated with Jacobs trucks and powered by a submarine engine. On its inaugural run between Denver and Chicago, the train averaged 77 mph (124 km/h) for the 1,015-mile (1,633-km) journey, and reached a top speed of 112¹/₂ mph (181 km/h).

▷ **SBB Doppelpfeil, 1939**

Wheel arrangement	2 x 4-wheel powered trucks (single unit)
Power supply	15 kV 17 Hz AC, catenary
Power rating	single units 528 hp (394 kW); twin units 1,126 hp (840 kW)
Top speed	77 mph (124 km/h)

Seven of the Schweizerische Bundesbahnen's _Rote Pfeil_ (or Red Arrow), streamlined electric single-unit railcars were introduced in 1935 for service on the Swiss Gotthard Railway, a major international rail link between Germany and Italy via the 49,222-ft (15,003-m) Gotthard Rail Tunnel. Three twin units known as _Doppelpfeil_ (or Double Arrows) were introduced in 1939.

Practical Diesels and Electrics

World War I had left Europe's railroads in tatters; coal was scant and expensive, and, while steam was still popular, other forms of traction would soon emerge to herald the end of an era. In mountainous countries such as Switzerland and Italy, an abundance of clean and cheap hydroelectric power made possible the electrification of main lines. Powerful electric locomotives, such as the Swiss "Krocodils," were soon hauling heavy trains over demanding routes, while in Italy speed records were being broken on Mussolini's new high-speed railroad. At the other end of the scale, small diesel and electric switchers (or shunters) were being introduced in both Europe and North America as a more efficient way of marshaling trains.

▷ **SBB Class Ce 6/8 II and Ce 6/8 III, 1919–20**

Wheel arrangement	1-C+C-1
Power supply	1.5 kV AC, catenary
Power rating	3,647 hp (2,721 kW)
Top speed	47 mph (76 km/h)

Serving until 1980, 51 of these electric engines were built to haul heavy freight on the Swiss Federal Railways' (Schweizerische Bundesbahnen, or SBB) Gotthard line from 1919 to 1927. Their long noses, for which they were nicknamed "Krokodils" (crocodiles), contain the motors.

△ **GIPR Class WCP 1, 1930**

Wheel arrangement	1'Co2'
Power supply	1.5 kV DC, catenary
Power rating	2,158 hp (1,610 kW)
Top speed	75 mph (121 km/h)

The first electric locomotives to be used in India, 22 of these powerful passenger engines were built from 1930 by Metropolitan-Vickers in the UK for the Great Indian Peninsula Railway. The first of these, No. 4006 *Sir Roger Lumley*, is on display at the National Rail Museum, New Delhi.

▷ **DR E04, 1933**

Wheel arrangement	1'Co1'
Power supply	15 kV AC 17 Hz, catenary
Power rating	2,694 hp (2,010 kW)
Top speed	75 mph (121 km/h)

A total of 23 Class E04 electric locomotives were built for Deutsche Reichsbahn for service on the newly electrified Stuttgart to Munich main line. Members of the class stayed in service in West Germany until 1976 and in East Germany until 1982. Several of these have been preserved.

▷ **GHE T1, 1933**

Wheel arrangement	A1 (0-2-2)
Transmission	mechanical
Engine	4-cylinder diesel
Total power output	92 kW (123 hp)
Top speed	25 mph (40 km/h)

This unique four-wheel 3-ft 3-in- (1-m-) gauge diesel railcar (*Triebwagen*) was built by Waggonfabrik Dessau in 1933 for the Gernrode-Harzgeroder Railway in Germany. After WWII it became No. 187.001 of the East German Deutsche Reichsbahn and was used as a workman's tool wagon. Seating 34 passengers, this restored railcar runs on the Harz narrow-gauge railroads.

△ PRR Class B1, 1934

Wheel arrangement	C (0-6-0)
Power supply	11 kV AC, catenary
Power rating	697 hp (520 kW)
Top speed	25 mph (40 km/h)

Fourteen of these single-unit electric switchers were built at Altoona Works by the Pennsylvania Railroad in 1934. They spent most of their life performing empty car movements in and out of Penn Station in New York City before retiring in the early 1970s.

△ DR Class Kö, 1934

Wheel arrangement	B (0-4-0)
Transmission	mechanical
Engine	79 hp (959 kW) diesel as modified
Total power output	24-29 hp (18-22 kW)
Top speed	11 mph (18 km/h)

These small diesel mechanical switchers, known as *Einheitskleinlokomotiven*, served at small stations on the Deutsche Reichsbahn. Equipped with only a foot brake, some were converted to run on LPG during WWII. Three of these, including No. 199.011 shown, have been converted to operate as Class Kö II on the 3-ft 3-in- (1-m-)gauge Harz railroads.

▷ LMS Diesel Shunter No. 1831, 1931

Wheel arrangement	C (0-6-0)
Transmission	hydraulic
Engine	Davey Paxman 6-cylinder diesel
Total power output	400 hp (298 kW)
Top speed	25 mph (40 km/h)

This was the first experimental diesel-hydraulic switcher in the UK. It was built by the London, Midland & Scottish Railway at its Derby Works in 1931 using the frame and running gear of a Midland Railway 1377 Class 0-6-0 steam locomotive of the same number. It was not successful and was officially withdrawn in 1939.

◁ FS Class ETR 200, 1937

Wheel arrangement	3-car articulated on 4-wheel trucks
Power supply	3 kV DC, catenary
Power rating	1,408 hp (1,050 kW)
Top speed	126 mph (203 km/h)

Entering service between Milan and Naples in 1937, a total of 18 of these three-car electric multiple units were built by Breda for the Italian state railroads. The streamlined shape was designed after wind tunnel tests, and in July 1939 unit ETR 212 set a world record for electric rail traction of 126 mph (203 km/h). The class was in regular service until the 1990s, and ETR 212 has since been preserved.

TECHNOLOGY

Track Inspection

During the 19th century, the maintenance of thousands of miles of train tracks, often in places inaccessible by road, was only made possible by teams of gangers walking the lines or traveling on unpowered handcars (also known as pump cars or jiggers). These were propelled by pushing a wooden arm up and down. By the 20th century, more ingenious methods had been introduced, such as motorized road vehicles with flanged wheels. Road-rail inspection vehicles are still used today on remote railroads around the world. In the US these are known as hi-rail vehicles; in Scotland, Land Rovers are adapted for use on the West Highland Line.

Buick Ma & Pa Car No. 101, 1937 Originally used as a funeral car, this vehicle was converted to run on the Maryland & Pennsylvania Railroad to test a radio communication system between locomotives and the railroad's offices.

Reading MU No. 800

The Reading Company (also known as the Reading Railroad) was a railroad and coal mining conglomerate that expanded rapidly from the 1830s. The company developed commuter rail services from Philadelphia, building the imposing Reading Terminal station in 1893. The decision to electrify many of the commuter lines was made in 1928 and, despite the Wall Street crash of 1929, the expansion continued and electric trains began running in July 1931.

THE READING MULTIPLE UNIT (MU) cars were specially built for the electrification project. Incorporating the latest technology, the cars were designed to be cheap to run. Aluminum was used extensively in the car bodies to make them light and reduce the amount of electricity needed to operate them. The MU cars were designed to work on their own or as part of much longer trains, as they had cabs at both ends. Furthermore, they were simpler to operate and much faster than the steam locomotives they had replaced.

The first 61 cars were ordered in 1928 and delivered in 1931. The company ordered more cars as the electric network continued to expand; by 1933, more than 84 miles (135 km) of the system was electrified. Some cars remained in service for 60 years, including 38 that were rebuilt between 1963 and 1965 and survived until 1990. Most of the older cars were withdrawn a year or two after the state government's Southeastern Pennsylvania Transportation Authority (SEPTA) took control of services on the former Reading Company lines in 1983.

FRONT VIEW

REAR VIEW

Electric and steam
The Reading Railroad used the "Reading Lines" brand name for its passenger services. In addition to running electric trains, the railroad operated several steam locomotives.

SPECIFICATIONS			
Class	EPa/EPb	In-service period	1931-90 (No. 800)
Wheel arrangement	B2	Railroad	Reading Railroad
Origin	United States	Power rating	480 hp (358 kW)
Designer/builder	Harlan & Hollingsworth	Power supply	11 kV AC 25 HZ overhead lines
Number produced	91 Reading MU cars	Top speed	70 mph (113 km/h)

Pantograph draws electrical power from overhead wires

Ventilator mounted on roof

Windows along full length of passenger car

Electric bus connector transfers power to next car

Electric bus connector
Visible above the headlight, the electric bus connector was first used in the US on the Reading Co. MU cars. The connectors on adjacent coaches touched, enabling electricity to pass safely from one car to the next. As a result, only a single pantograph was required to operate several motor cars at once.

EXTERIOR

Looking similar to today's commuter rail cars, the MU car—with its steel body sides, driving cabs at either end, and automatic doors—was technologically advanced when introduced. High-voltage power lines ran along the vehicle roof from the pantograph (which made contact with the overhead wire power supply) to the electric bus connectors at each end of the car.

1. Railcar number on side **2.** Electric bus connector at each end of car **3.** Headlight **4.** Horn **5.** Insulator for 11 kV AC electric power cable running along roof **6.** Pantograph in lowered position **7.** Marker light **8.** Cab window and wiper blade **9.** Cowcatcher **10.** Sockets for 12-pin multiple working cables **11.** 12-pin multiple working cables **12.** Brake shoe **13.** Open journal box showing bearing **14.** Leaf spring suspension on wheel unit **15.** *Taylor Flexible Truck* **16.** Handbrake chain **17.** Air brake control valves

CAB INTERIOR

The engineer of a Reading Company MU car had to stand up in the cab, using simple controls developed from those used on electric tramways. They also had to estimate the train's speed, as the early trains did not have speedometers. However, the MU was equipped with a cab signaling system that delivered electrical signals to the train via the track.

18. Engineer's cab   **19.** Handbrake   **20.** Ratchet for handbrake
21. Throttle control, with socket for operation by Allen wrench
22. Light switch boxes (left) and brake pressure gauges   **23.** Marker light lens (red) indicates end of train   **24.** Marker light lens (yellow, shows as white when lit by oil lamp) indicates unscheduled train   **25.** Marker light lens (blue, shows as green when lit by oil lamp) indicates scheduled train
26. Top of brake unit   **27.** Cab interior door lock

CAR INTERIOR

The standard MU car had 18 rows of 2+2 reversible seats with space for gangway-facing seats and a conductor's office. The freight cars had a separate compartment at one end and fewer seats. The railroad also introduced innovations such as internal doors that automatically closed when the train was in motion and thermostats for heating, which reduced costs and improved passenger comfort.

28. Interior of railcar   **29.** Light fixture in ceiling   **30.** Metal luggage rack above seats
31. Brass window sash clip   **32.** Footrest and heater under seat   **33.** Door to washroom
34. Emergency ax in glass case

1940-1959
WAR AND PEACE

WAR TRAFFIC MUST COME FIRST

FRED CHANCE

DON'T WASTE TRANSPORTATION

ASSOCIATION OF AMERICAN RAILROADS IN COOPERATION WITH THE OFFICE OF DEFENSE TRANSPORTATION

WAR AND PEACE

During World War II railroads formed an integral part of the struggle for victory for both Allied and Axis forces. However, rail traffic not directly related to the war effort was discouraged during this period. Many of Europe's railroads were devastated during the conflict, and they were also much maligned by association for transporting millions of people to concentration camps. But emerging from the shadows of war, Western Europe's railroads returned with a new glamorous face, the Trans-Europ Express (TEE)—part of a major international effort to rebuild and rebrand railroads in the war's aftermath.

Although the US had rolled out steam giants such as the "Big Boy" in the early 1940s, the shift to diesel had already started with a rapid transition to the new form of traction. During the postwar period throughout the world, diesel and electric power increasingly replaced steam, which was seen as dirty, labor-intensive, and old-fashioned. The supremacy of the new technologies was enhanced in 1955 when two French electric trains broke the world speed record.

In the UK, the newly nationalized British Railways stuck with steam power until after the publication of the Modernization Plan in the mid-1950s. Nevertheless, the ever-increasing use of diesel switching engines showed the way forward, and in 1955 the prototype *Deltic* appeared, presenting the new face of express travel. However, by the late 1950s, neither the US nor Europe was developing the most revolutionary form of rail transportation. That honor went to Japan.

△ **Red carpet train**
A 1941 poster advertises the famous *20th Century Limited* passenger train, which later featured in the film *North By Northwest*.

> "The railroads ... can be **reached at any moment** by military orders. **Nothing**, therefore, **can replace the railroads**"
> ERNST MARQUARDT, GERMAN MINISTRY OF TRANSPORTATION, 1939

◁ **Propaganda poster (1939-45) by Fred Chance,** who worked as an illustrator in Philadelphia and New York

Key Events

▷ **1941** In the United States, the Union Pacific Railroad launches its giant "Big Boy" steam locomotives.

▷ **1942** Germany's Class 52 "Kriegslok" (war loco) is introduced as a stripped-down wartime design. Its reliability also helps the postwar reconstruction.

▷ **1945** Allied forces use "train-busters" to destroy German locomotives.

△ **Casualty of war**
This German locomotive was found upended by Allied forces when they captured the town of Muenster, Germany on April 11, 1945.

▷ **1948** Railroads are nationalized in the UK. Private companies are replaced by British Railways.

▷ **1949** With the formation of West Germany, the country's railways become the Deutsche Bundesbahn. East Germany's railroads keep the name Deutsche Reichsbahn.

▷ **1951** British Railways launches a new range of "Standard" steam designs.

▷ **1954** December 1, the British Railways Modernization Plan announces the elimination of steam.

▷ **1955** French electrics BB 9004 and CC 7107 reach 206 mph (331 km/h)— a new world record.

▷ **1957** The pan-European Trans-Europ Express (TEE) network is launched, and a series of iconic trains are built to run on its routes.

World War II Logistics

The transportation of raw materials, troops, military equipment, and ammunition by rail was of strategic importance to the warring powers in World War II. As a result, cheaply constructed, powerful freight locomotives—mass-produced in Germany, Britain, and the US— saw active service in war zones. After the war, many ran on European national railroads, as replacements or as war reparations. A large number of engines, built for the United States Army Transportation Corps (USATC), were sent to Asia under lease-lend agreements and, after the war, by the UN Relief & Rehabilitation Administration.

◁ LMS 8F, 1935

Wheel arrangement	2-8-0
Cylinders	2
Boiler pressure	225 psi (15.82 kg/sq cm)
Drive wheel diameter	56¼ in (1,430 mm)
Top speed	approx. 50 mph (80 km/h)

Designed by William Stanier for the London Midland & Scottish Railway, these were the standard British freight locomotives for part of WWII. They saw service for Britain's War Department in Egypt, Palestine, Iran, and Italy—25 were sold to Turkey in 1941. Of the 852 built, some remained in British service until 1968, while Turkish examples ran into the 1980s.

△ DR Class 52 "Kriegslok," 1942

Wheel arrangement	2-10-0
Cylinders	2
Boiler pressure	232 psi (16.3 kg/sq cm)
Drive wheel diameter	55 in (1,400 mm)
Top Speed	50 mph (80 km/h)

Around 7,000 of these Deutsche Reichsbahn heavy freight locomotives were built mainly for service on the Eastern Front. A small number remain in service in Bosnia even today, while many, like this Class 52 No. 52.8184-5 rebuild, have been preserved.

◁ USATC S160, 1942

Wheel arrangement	2-8-0
Cylinders	2
Boiler pressure	225 psi (15.82 kg/sq cm)
Drive wheel diameter	56³/₄ in (1,440 mm)
Top speed	approx. 45 mph (72 km/h)

Of the 2,120 austerity Consolidation-type heavy freight locomotives built for the USATC, 800 were shipped to Britain for use in Europe after D-Day. After the war, they saw service on many European railroads as well as in North Africa, China, India, and North and South Korea.

▷ USATC S100, 1942

Wheel arrangement	0-6-0T
Cylinders	2
Boiler pressure	210 psi (14.8 kg/sq cm)
Drive wheel diameter	54 in (1,372 mm)
Top speed	approx. 35 mph (56 km/h)

Built for the USATC, 382 of these locomotives were shipped to Britain and used in Europe after the D-day landings of June 1944. Britain's Southern Railway later bought 15 as shunters.

△ Class V36 Shunter, 1937

Wheel arrangement	0-6-0
Transmission	Hydraulic
Engine	Deutsche Werke/MAK diesel
Total power output	360 hp (268 kW)
Top speed	approx. 37 mph (60 km/h)

Equipped with four axles but only three pairs of drive wheels, these diesel locomotives were built for the German armed forces (Wehrmacht), and were used for shunting duties. They saw widespread use in Europe and North Africa after the war.

▷ Armored Car, 1942

Type	4-wheel
Capacity	130 (whole train)
Construction	Armor-plated steel
Railroad	German Wehrmacht

This camouflaged car formed part of a German Wehrmacht BP42 armored train that protected supply and transport trains in the Balkans and Russia. An armored Class 57 0-10-0 steam locomotive was positioned in the center of the train, which consisted of a combination of infantry, navigating, anti-aircraft, and artillery wagons, with converted tank turrets.

TALKING POINT

△ SR Class Q1, 1942

Wheel arrangement	0-6-0
Cylinders	2 (inside)
Boiler pressure	230 psi (16.17 kg/sq cm)
Drive wheel diameter	61 in (1,549 mm)
Top Speed	50 mph (80 km/h)

Designed by Oliver Bulleid for the Southern Railway, these freight locomotives were lightweight, which enabled them to operate over most of the company's network. A total of 40 were built, and they all remained in service on the Southern Region of British Railways until the 1960s. This is No. C1, the first of the series.

The Maryland Car

In 1947, US journalist Drew Pearson set out to help the people of war-stricken France and Italy. A Friendship Train traveled around the US gathering $40 million of relief supplies. In response, the French sent a Merci (thank you) Train filled with gifts back to the US. Arriving in New York in 1949, the train consisted of a series of European boxcars used to transport soldiers and horses during the war. There were 49 cars—one for each of the then 48 US states plus another for the District of Columbia and Hawaii to share. The Maryland Car, shown here, was originally built for the Paris, Lyon & Mediterranean Railroad in 1915. It is now on display at the Baltimore & Ohio Railroad Museum, Baltimore.

△ WD Austerity, 1943

Wheel arrangement	2-8-0
Cylinders	2
Boiler pressure	225 psi (15.82 kg/sq cm)
Drive wheel diameter	56$\frac{1}{4}$ in (1,430 mm)
Top speed	approx. 45 mph (72 km/h)

Designed by R. A. Riddles for the British War Department, these heavy freight trains were "austerity," or cheaper, versions of the LMS 8F. Of the 935 built, many saw service in mainland Europe after D-day in June 1944. After the war, 733 were in operation for British Railways, while others worked in the Netherlands, Hong Kong, and Sweden.

▽ IR Class AWE, 1943

Wheel arrangement	2-8-2
Cylinders	2
Boiler pressure	210 psi (14.76 kg/sq cm)
Drive wheel diameter	61$\frac{1}{2}$ in (1,562 mm)
Top speed	approx. 62 mph (100 km/h)

These huge locomotives were built by Baldwin Locomotive Works for the USATC for hauling heavy freight trains in India during WWII. They had 7-ft (2,134-mm) diameter boilers, and 40 became Indian Railways Class AWE. One of these, No. 22907 *Virat*, has been restored to working order at the Rewari Steam Loco Shed.

DR No. 52.8184-5

Built to serve Germany during World War II, the Deutsche Reichsbahn (DR) Class 52 "Kriegslok" had a simple design and was constructed from materials that were easy to source during wartime. Nevertheless, it became a rugged classic vital to many countries long after the conflict ended, partly because it could haul heavy loads on lightweight tracks. Although designed to last only a few years, the class also proved very durable.

THE CLASS 52 "KRIEGSLOK" (*Kriegs-Dampflokomotive,* or war steam locomotive) came out of Germany's need to construct locomotives quickly during World War II, while maintaining maximum production capacity for armaments. The plan was to build 15,000 locomotives, with production spread throughout occupied Europe. Only around 7,000 were actually made, but Germany's Class 52 remains one of the most numerous classes ever built. No.52.8184-5 was built in Vienna in 1944.

The ten drive wheels gave the 52 enough grip to pull 2,000 tons across a level surface at 31 mph (50 km/h). In addition, the locomotive included plenty of cold-weather protection, useful in a war that progressed into Russia in winter. Some were even equipped with tenders that could recycle exhaust steam back into water, meaning they could travel long distances without having to refill.

After the war, "Kriegsloks" remained in service. Some were modernized, including the engine now known as No.52.8184-5, which is kept in Stassfurt, Germany.

FRONT VIEW   **REAR VIEW**

Deutsche Reichsbahn

Manufacturing for war
Like much of German industry during World War II, the Deutsche Reichsbahn was harnessed to the war effort. The Class 52 "Kriegslok" epitomised the machines of war built during that period.

SPECIFICATIONS			
Class	52 or Kriegs-Dampflokomotive 1	**In-service period**	1942–present (No.52.8184-5)
Wheel arrangement	2-10-0	**Cylinders**	2
Origin	Germany	**Boiler pressure**	232 psi (16.3 kg/sq cm)
Designer/builder	Hauptausschuss Schienenfahrzeuge	**Drive wheel diameter**	55 in (1,400 mm)
Number produced	approx. 7,000 Class 52	**Top speed**	50 mph (80 km/h)

Smoke deflectors keep exhaust away from driver's view

Steam dome collects steam to be used from the boiler

Cab is fully enclosed to protect the crew in cold conditions

Coal space is narrow, allowing engineer a clear reverse view

Water tank is frameless, cutting production costs

Simple elegance
Paring the design to the bare minimum helped give the Class 52 its austere look. Essential components were simplified wherever possible.

52 8184-5

H. Unt
03.04
V

EXTERIOR

Looks were not a priority for the designers of the wartime Class 52, though later refinements to the design softened the austere appearance of some engines. With functionality as the main goal, the locomotives were simply built using readily available materials, and key parts were made easily accessible. Much of the look was determined by the small drive wheels, which gave pulling power and low axle weight rather than high speed. Another distinctive aspect of many "Kreigsloks" was a tub-shaped tender.

1. Number plate on front of engine **2.** Smokebox door handle **3.** Front headlight **4.** Front buffer **5.** Stack **6.** Crosshead assembly **7.** Shut-off valve **8.** Inspection lamp under running plate **9.** Air pump **10.** Wheel unit with connection rod **11.** Air tank **12.** Door and window of cab; smaller windows meant that locomotive was less likely to be seen by World War II bombers **13.** Cab steps **14.** Leaf spring suspension on tender truck **15.** Tender tank **16.** Top lamp and number plate on rear of tender

1 52 8184-5

CAB INTERIOR

Built to keep out the winter cold, the "Kriegslok" cab was austere and simple, but all the instruments were positioned for ease of use. Following the standard German layout, the engineer sat on the right and the fireman worked from the left. The regulator handle, reverser, and brake controls were all within easy reach of the engineer, while the fireman had access to controls to regulate how much water he fed into the boiler. The firebox door swung on a flap and featured handles on both sides so the engineer could open it while the fireman shoveled the coal in.

17. Overview of cab interior **18.** Firehole
19. Inside of firebox **20.** Controls on fireman's side **21.** Lubricator **22.** Water-gauge glass
23. Sanding controls **24.** Air brake pressure gauge **25.** Sign explaining the correct use of Trofimoff valves **26.** Reverser **27.** Brake controls **28.** Switches for lamps **29.** Access to tender

Wartime Service

The importance of railroads in World War II can be measured by the efforts made by both sides to destroy them. Between 1940 and 1942, Germany's air force, the Luftwaffe, launched more than 10,000 attacks on Britain's rail network, but failed to prevent it from transporting the fuel, food, equipment, and munitions that the nation needed. Germany's railroads not only moved vital goods but also played a part in history's worst act of genocide—the transportation of Jews and other groups to death camps.

On the Allied side, the Trans-Iranian Railway in the Middle East kept oil flowing from the Persian Gulf to the Soviet Union, while in North America, trains delivered supplies to Atlantic ports for shipment to Britain.

THE BIG BUILD

By 1942, with the tide turning in the Allies' favor, planning began for the invasion of Axis-occupied Europe. Recognizing that railroads would be crucial to success, Britain and the US began a program of locomotive building on an unprecedented scale. By D-Day, June 6, 1944, more than a thousand new engines had been built to haul supply trains, and they were immediately put into service. The Allies achieved victory in May 1945 and, while there were many factors in their success, there is no doubt that railroads played a critical part.

Women filled roles left vacant by the railroadmen serving in World War II. Here, members of the British Women's Voluntary Service (WVS) clean the engines at a London Midland & Scottish Railway depot.

The US Moves into Diesel

The diesel locomotive represented the greatest advance in US railroading during the 20th century. Although diesel engines had been around since the 1890s, the challenge was to make them small and light enough to fit within the confines of a locomotive, yet powerful enough to haul a train. The breakthrough came in 1935, when General Motors unveiled a 12-cylinder, 2-cycle engine that was 23 percent smaller and, thanks to lightweight alloys, 20 percent lighter than its predecessors. Accelerating into the 1940s, the diesel began to conquer the US.

◁ **Boxley Whitcomb 30-DM-31, 1941**

Wheel arrangement	0-4-0
Transmission	mechanical
Engine	8-cylinder Cummins
Total power output	150 hp (112 kW)
Top speed	approx. 20 mph (32 km/h)

Built by the Whitcomb company of Rochelle, Illinois, "30" referred to the locomotive weight range in tons and "DM" to its transmission (diesel-mechanical). The Boxley Materials Co. of Roanoke, VA bought No. 31 from the Houston Shipbuilding Corporation of Texas in 1953.

◁ **VC Porter No. 3, 1944**

Wheel arrangement	A1-A1
Transmission	electric
Engine	not known
Total power output	300 hp (224 kW)
Top speed	approx. 20 mph (32 km/h)

H.K. Porter Inc. was one of the largest manufacturers of industrial locomotives in the US—by 1950, it had built 8,000. This rod-driven switcher Porter No. 3, built for the Virginia Central Railroad, is the last of the 28 of its type built. It is now preserved at the Virginia Museum of Transportation.

▷ PMR GM EMD SW-1 No. 11, 1942

Wheel arrangement Bo-Bo

Transmission electric

Engine EMD Model 567 V-6 engine

Total power output 600 hp (448 kW)

Top speed 45 mph (72 km/h)

The Pere Marquette Railway, which took its name from a 17th-century French Jesuit priest, served the Great Lakes region of the US and Canada. The EMD SW-1 Class was introduced in 1936, but No. 11 was delivered to the railroad in April 1942 to begin switching hoppers at Eireau, Ontario. It was retired in 1984.

△ **GM Class E7a, 1945**

Wheel arrangement A1A-A1A

Transmission electric

Engine 2 x EMD Model 567A, 12-cylinder

Total power output 2,000 hp (1491 kW)

Top speed 85 mph (137 km/h)

Supplied to over 20 railroads, General Motors's E7 was one of the first standard American diesels. Between 1945 and 1949, 428 of the E7a cab units were produced along with 82 E7b boosters. The Gulf, Mobile, & Ohio Railroad's No. 103, shown here, was notable for appearing in the movie *The Heat of the Night*.

△ **Baldwin Class DS-4-4-660, 1946**

Wheel arrangement Bo-Bo

Transmission electric

Engine 4-cycle engine

Total power output 660 hp (492 kW)

Top speed 60 mph (96 km/h)

The Chesapeake & Western (known locally as the "Crooked and Weedy") operated over 53½ miles (86 km) of the Shenandoah Valley. In 1946, it took delivery of three diesel units. With running costs at 25 cents per mile (as opposed to the 96 cents for steam), it marked a turning point for the CHW. No. 662 was retired in 1964 and languished in a scrapyard before being donated to the Virginia Museum of Transportation.

◁ **Ma&Pa GM EMD Type NW2, 1946**

Wheel arrangement Bo-Bo

Transmission electric

Engine 12-cylinder engine

Total power output 1,000 hp (750 kW)

Top speed 60 mph (97 km/h)

This type was first introduced in 1939. The Maryland & Pennsylvania Railroad (a short line linking Baltimore with York and Hanover, PA) took delivery of Nos. 80 and 81, two Type NW2 switchers built by the Electro-Motive Division of General Motors, in December 1946. In total, 1,145 NW2s were shipped between 1939 and 1949 to over 50 railroads (in contrast to the Ma&Pa's modest pair, Union Pacific bought 95). No. 81 has been part of the Railroad Museum of Pennsylvania's collection since 1997.

Postwar US

Some railroads needed persuading that the diesel could match the hauling power of steam, but once General Motors' freight demonstrator and prototypes had convinced them, the economic argument was irresistible. Mainline locomotives fell into two categories: cab units and hood units. The former, with their sleek bodywork and colorful liveries, handled the expresses and were augmented by boosters for extra power. In the hood unit, the workhorse, the engine (or engines), radiators, and ancillary equipment were mounted on a platform above the chassis with the cab placed at one end or in the center. The transition from steam to diesel was accomplished within 20 years; by 1960 around 34,000 diesel locomotives operated in the US.

▷ B&A GE 70-ton switcher, 1946

Wheel arrangement Bo-Bo

Transmission electric

Engine 2 x Cooper-Bessemer FDL-6T 6-cylinder 4-cycle engines

Total power output 660 hp (492 kW)

Top speed 60 mph (96 km/h)

The Baltimore & Annapolis Railroad was mainly a commuter line that in 1950 succumbed to road competition and replaced passenger trains with buses. That year it bought a solitary diesel, a General Electric 70-ton switcher— No. 50—for freight operations. The type was introduced in 1946 as a lighter, low-cost option for secondary routes, and 238 were built up to 1955. Retired in 1986, No. 50 is preserved at the Baltimore & Ohio Museum.

△ Baldwin S12 switcher, 1950

Wheel arrangement Bo-Bo

Transmission electric

Engine De Lavergne Model 606A SC 4-cycle engine

Total power output 1,200 hp (895 kW)

Top speed 60 mph (96 km/h)

Employing a turbocharged version of the powerful Model 606A engine, the S12 switcher was famous for its hauling prowess, as demonstrated by Baldwin's original No. 1200. Here, masquerading as No. 1200, is Earle No. 7 or, in the records of its operators, the United States Navy, No. 65-000369. The USN took 18 of the 451 S12s shipped between 1951 and 1956 and stationed this unit at its ordnance depot in Earle, NJ.

▷ B&O F7 Class, 1949

Wheel arrangement Bo-Bo

Transmission electric

Engine EMD 567B 16-cylinder engine

Total power output 1,500 hp (1,119 kW)

Top speed approx. 50–120 mph (80–193 km/h)

The F7 was the most numerous of the General Motors F Series; 2,341 A units and 1,467 B (booster) units were built by 1953. The speed variation is a product of eight different gear ratios. Though tailored for freight, many railroads used F7s for front-line passenger services until the 1970s. No. 7100, shown, was bought by the Baltimore & Ohio Railroad in 1951 and enjoyed a second career on the Maryland Area Regional Commuter (MARC) system from 1987 to the late 1990s.

◁ N&W EMD GP9 Class, 1954

Wheel arrangement Bo-Bo

Transmission electric

Engine EMD 567C 16-cylinder engine

Total power output 1,750 hp (1,305 kW)

Top speed 75 mph (121 km/h)

General Motors' "Geep Nine" remains one of the most successful and long-lasting of diesels, although not the most attractive. Looks did not count for US and Canadian railroads, which between them bought 4,087 A units and 165 type B boosters from 1954 to 1963. No. 521 was one of 306 GP9s on the books of the Norfolk & Western, and many remain in service on secondary lines and with industrial users; some Class 1 railroads still use them as switchers.

△ Budd RDC railcar, 1949

Wheel arrangement	Bo-2
Transmission	mechanical
Engine	2 x General Motors Type 6-110 6-cylinder engines
Total power output	275 hp (205 kW)
Top speed	85 mph (137 km/h)

After WWII, the Budd Co. used its expertise in building lightweight, stainless-steel train cars to assemble diesel railcars (or multiple units) for secondary and local passenger services. A prototype Rail Diesel Car (RDC) was unveiled in 1949 and offered impressive economy. By 1962, 398 were in operation. Out west, RDCs provided a stopping service over the 924 miles (1,487 km) between Salt Lake City, UT, and Oakland, CA. The RDC was also exported to Australia, Brazil, Canada, and France.

◁ N&W ALCO T6 (DL440) Class, 1958

Wheel arrangement	Bo-Bo
Transmission	electric
Engine	ALCO 251B 6-cylinder 4-cycle engine
Total power output	1,000 hp (746 kW)
Top speed	60 mph (96 km/h)

The American Locomotive Co. (ALCO) introduced the T6 (the "T" stood for "transfer") in 1958, believing there was a demand for a switcher capable of shuttling trains between yards and terminals at higher speeds. This was not the case, and up to 1969 just 57 had been delivered, of which the Norfolk & Western Railway took 38. Retired in 1985, No. 41 is kept at the Virginia Museum of Transportation.

N&W GP9 Class No. 521

The last major US rail operator to switch from steam to diesel was the Norfolk and Western Railway (N&W), based in Roanoke, VA. As part of its drive to eliminate steam, N&W already ran electric trains on some of its routes. From 1955 it moved to diesel, first with ALCO RS-3s, and then bought 306 model GP9 diesel-electric locomotives from General Motors's Electro-Motive Division (EMD). Most of the GP9s were destined for freight duties but a small batch, including No. 521, hauled passenger trains.

THE EMD GENERAL PURPOSE (GP) road switcher diesel engines first appeared in 1949 and became the most successful range of mid-power diesels in North America. The first model, the GP7, was built from 1949 to 1954, when the improved GP9 version was introduced. The "Geeps," as the locomotives were nicknamed, were bought in large numbers to replace the steam locomotives still in use during the 1950s. Continuously updated, the last "Geep" model was the GP60, produced until 1994.

Locomotives 501 to 521 were the last GP9s bought by N&W and were equipped with steam boilers to heat passenger coaches. At first they replaced the fast J Class steam locomotives that worked the N&W passenger trains in the 1950s, but when the passenger services ceased they were used for freight alongside the 285 other GP9s operated by this railroad. In 1982, N&W merged with the Southern Railway to become the Norfolk Southern Railway, which today is one of the largest Class 1 railroads in the United States.

FRONT VIEW

REAR VIEW

Special logo
For the introduction of the new passenger GP9 diesels the N&W logo in yellow was, unusually, mounted on a round plate with a black background on the locomotive front.

SPECIFICATIONS			
Class	GP9	**In-service period**	1958–85 (No. 521)
Wheel arrangement	BoBo	**Transmission**	electric
Origin	United States	**Engine**	EMD 567C 16-cylinder
Designer/builder	General Motors EMD	**Power output**	1,750 hp (1,305 kW)
Number produced	306 (GP9s for N&W)	**Top speed**	75 mph (125 km/h)

Tuscan red livery given to N&W passenger locomotives

Dynamic brake grille dissipates heat from brakes

Fuel tank could hold 1,080 gallons (4,090 liters) of diesel

Twin air horn mounted on driver's cab roof

Safety railings run full length of locomotive

Brass bell used to alert staff and passengers when moving in yard or station

NORFOLK AND WESTERN

521

"The redbirds"
The last 21 GP9s bought by N&W for passenger trains were given a special livery of Tuscan red with yellow lettering, earning them the nickname "the redbirds."

EXTERIOR

The GP9 was a simple but rugged design with features in common with all EMD locomotives of the time, such as the standard US "knuckle" coupler, originally developed in the 1890s and installed at each end of the GP9. The use of spare parts that were interchangeable between EMD models was one of the reasons so many of these locomotives were sold in the 1950s.

1. Number plate on front end of engine 2. Twin headlights 3. Ladder to access top of engine 4. Electrical connection cap 5. "Knuckle" coupler
6. Diesel fuel cap 7. Emergency fuel cutoff 8. Front steps 9. Wheel unit (truck) 10. Air horn positioned above cab 11. Spring on engine truck
12. Air-brake cylinder 13. Dynamic brake grille 14. Clasp brake
15. Brass bell on front end 16. Door to cab

CAB INTERIOR

The engineer's cab had a standard EMD control station, with lever-operated power and reverse and braking controls. The locomotive and train brake equipment were located alongside each other. The locomotive could be driven in either direction at full speed, and the engineer had a good view forward from each end of the cab. The power controller (or throttle) had eight "notches," so the engineer could increase or decrease power gradually.

17. Interior of cab with engineer's controls **18.** Emergency brake valve **19.** Windshield wiper motor **20.** Switches for windshield wipers **21.** Brake control levers **22.** Warning sign **23.** Speedometer **24.** Control panel circuit breaker switches **25.** Air brake gauges **26.** Load indicator **27.** Power controller

Great Britain Makes the Change

By the 1940s, the British rail system consisted of four major companies and many smaller light railroads. In 1948 the "Big Four" and the majority of the smaller railroads were nationalized under one umbrella company, British Railways. The new company commissioned a report to look at ways of stemming the losses they were incurring as a result of competition from air and road traffic. Known as the Modernization Plan and published on December 1, 1954, the report made a number of recommendations, including the replacement of all steam engines. Tests in the late 1950s with "pilot-scheme" diesels were intended to discover which locomotives to order in quantity. Orders for thousands of new diesels would follow in the next decade.

△ BR (W) Gas Turbine No. 18000, 1949

Wheel arrangement	A1A-A1A
Transmission	electric
Engine	Brown Boveri Gas Turbine
Total power output	2,500 hp (1,865 kW)
Top speed	90 mph (145 km/h)

This revolutionary locomotive was delivered to British Railways in 1949 from Switzerland and was used for 10 years on the BR Western Region. In 1965 it left the UK and was used for research in Switzerland and Austria, returning in 1994 to the UK, where it is now preserved.

△ BR Class 08, 1953

Wheel arrangement	0-6-0
Transmission	electric
Engine	English Electric 6KT
Total power output	350 hp (261 kW)
Top speed	20 mph (32 km/h)

Based on a wartime-design diesel switcher ordered by the London, Midland & Scottish Railway, over 950 Class 08 locomotives were built by five British Railways workshops between 1953 and 1959. Smaller batches of similar locomotives using different engines were also built. Sixty years on, some remain in service. No. 08 604 *Phantom* is preserved.

△ BR Class 05, 1954

Wheel arrangement	0-6-0
Transmission	mechanical
Engine	Gardner 8L3
Total power output	201 hp (150 kW)
Top speed	17 mph (27 km/h)

This engine was one of several designs of smaller switching locomotives delivered to British Railways in the 1950s. Later classified as Class 05, 69 were built between 1954 and 1961. Few remained in service for more than a decade, as the freight traffic they were built for disappeared after the BR network was reduced following the Beeching Report.

◁ English Electric prototype *Deltic*, 1955

Wheel arrangement	Co-Co
Transmission	electric
Engine	2 x Napier Deltic D18-25 engines
Total power output	3,300 hp (2,461 kW)
Top speed	106 mph (171 km/h)

Built speculatively by English Electric, *Deltic* was the prototype for the 22 Type 5 *Deltic* D9000 Class 55 diesel locomotives bought for services on the East Coast route from London to York and Edinburgh. They were to replace the famous London & North Eastern Railway design A4 Pacific steam engines.

18000

D 2595

▷ **BR Type 1 Class 20, 1957**

Wheel arrangement Bo-Bo

Transmission electric

Engine English Electric 8SVT MkII

Total power output 986 hp (735 kW)

Top speed 75 mph (121 km/h)

This class was one of the most successful of all the Modernization Plan locomotives. A total of 227 were built for British Railways between 1957 and 1968. The class saw limited passenger services but could work in multiples and, coupled together, could handle heavy traffic. Some remain in use with UK freight operators nearly 60 years later.

D 801

◁ **BR Class 42, 1958**

Wheel arrangement B-B

Transmission hydraulic

Engine 2 x Maybach MD650 engines

Total power output 2,100 hp (1,566 kW)

Top speed 90 mph (145 km/h)

These locomotives were based on successful V200 engines that ran in West Germany and used the same engines as their German cousins. Known as *Warships*, they were used by British Railways principally on the Western Region from London Paddington to Devon, Cornwall, and South Wales until withdrawn from service in 1972. This is No. 801 *Vanguard*.

△ **BR Class 108, 1958**

Wheel arrangement 2-coach multiple unit

Transmission mechanical

Engine 2 x BUT/Leyland 6 cylinder

Total power output 300 hp (224 kW)

Top speed 70 mph (113 km/h)

British Railways's Modernization Plan led to the replacement of steam locomotives, and more than 4,000 diesel multiple units were ordered. These new self-propelled "Derby Lightweight" trains were much cheaper to operate than the steam trains they replaced.

▷ **BR Type 4 Class 40, 1958**

Wheel arrangement 1Co-Co1

Transmission electric

Engine English Electric 16SVT MkII

Total power output 1,972 hp (1,471 kW)

Top speed 90 mph (145 km/h))

This class was designed to replace the fastest steam locomotives working express trains initially between London and Norwich and later all over the UK. The initial pilot batch of 10 was expanded to a final class of 200 by 1962.

Deltic Prototype

During its time, the English Electric prototype *Deltic*, first tested in 1955, was the most powerful diesel locomotive in the world. Using Napier Deltic engines developed to power fast naval patrol boats, *Deltic* produced high levels of performance while weighing less than most contemporary locomotives. British Railways ordered 22 production *Deltics* in 1958, introducing 100 mph (161 km/h) express trains to the UK.

THE NAPIER DELTIC ENGINE had a unique layout with three banks of six cylinders in a triangular formation. To enable each group of cylinders to work efficiently, the crankshaft for one group operated in the opposite direction of the other two; the resulting opposed piston engine was both compact and very powerful. Its light weight meant that two derated naval engines could be installed in a six-axle locomotive; the power available made the *Deltic* the most powerful diesel locomotive of its time.

The *Deltic* prototype began tests with British Railways in 1955, initially on the West Coast route from London to Liverpool and Carlisle. The locomotive remained the property of its builders, English Electric, whose engineers accompanied it on every trip. From 1959 it operated on the East Coast route from London to York and Edinburgh. It was here that it would excel, leading to an order for 22 locomotives, with a slightly smaller body shell, which would replace 55 express passenger steam engines. Retired from use in 1961, the prototype was presented to London's Science Museum and today is part of the UK's National Railway Museum collection.

SIDE VIEW

FRONT VIEW

Major manufacturer
The Deltic prototype was built and owned by English Electric. Founded in 1918 this major British engineering company built hundreds of diesel and electric engines until 1968.

SPECIFICATIONS	
Class	Deltic prototype
Wheel arrangement	Co-Co
Origin	UK
Designer/builder	English Electric, Vulcan Foundry
Number produced	1
In-service period	1955-61
Transmission	Electric
Engine	2 x Napier Deltic D18-25
Power output	3,300 hp (2,460 kW)
Top speed	106 mph (171 km/h)

A spacious cab was provided at either end of the locomotive

Engine room housing the two Napier Deltic engines and two generators

Nameplate is unusual, as the prototype had no number, just the name

Bright blue, cream, and gold livery is unique to the prototype and has not been seen on a British locomotive before or since

American appeal
The headlight was part of the rounded North American styling for the prototype locomotive. In practice, this headlight was never used because the locomotive never left the UK for trials anywhere else.

EXTERIOR

The *Deltic* prototype was built with export markets around the world in mind and consequently used styling similar to US streamlined diesel designs that had been in use since the 1940s. The bright blue, cream, and gold livery made it stand out from every other locomotive in the UK when it started trials in 1955. The design had many innovative features for its era, such as retractable steps, streamlined lights, and buffers.

1. Painted name plaque 2. Large headlight space (light never installed) 3. Streamlined electric marker light 4. Front buffer 5. Front coupling hook 6. Horn bracket 7. Windshield and wiper blades 8. Sandbox 9. Folding chrome steps 10. Air brake chain 11. Exhaust vent positioned at center of engine 12. Metal steps up to driver's door 13. Leaf spring suspension 14. Fuel gauge 15. Shed shore supply (electricity)

ENGINE ROOM

The design for the compact Deltic engine originated in World War II Junkers airplane engines from Germany. The engine is made from aluminum alloy and designed to be as lightweight as possible. Fitting the engines into the locomotive was an engineering challenge, as the loading gauge (maximum height and width) of UK trains is smaller than on European rail systems.

16. Napier Deltic Engine 17. Controls at top of engine 18. Steam heating boiler

CAB INTERIOR

The locomotive had identical cabs at either end, each designed to give a good view forward, from a raised position, through a two-piece windshield. This clear view of the line ahead was essential for safe operation at a speed of 100 mph (161 km/h). The locomotive was operated by a two-man crew, one driving and the other monitoring ancillary equipment, such as steam heating.

19. Left side of cab **20.** Right side of cab **21.** Warning light attached to ceiling **22.** Westinghouse vacuum brake **23.** Wiper motor **24.** Driver's display panel **25.** Loco brake (above) and power handle (below) **26.** Orange electricity conduits **27.** Maintenance doorway to the nose **28.** Vacuum exhauster in the nose **29.** Wheel brake **30.** Steam heating control

Europe Follows the US

As Europe emerged from the chaos and damage inflicted by World War II, many railroad companies based their future planning on the US, where diesels had been replacing steam for nearly a decade. A wide variety of manufacturers using an equally wide choice of diesel engines built locomotives for state railroads across Europe. Labor-intensive steam was replaced with diesels, which were cheaper to run, although more expensive to buy. The process was gradual in most countries; some steam engines survived until 1977 in West Germany, and they never entirely disappeared in East Germany.

▽ DB V200 (Class 220), 1954

Wheel arrangement	B-B
Transmission	hydraulic
Engine	2 x Maybach MD 650 engines
Total power output	2,170 hp (1,618 kW)
Top speed	87 mph (140 km/h)

Designed to replace steam locomotives on heavy express passenger trains in the mid-1950s, the Class 220s were displaced to less important routes by electrification in the 1960s and 70s. All were withdrawn by Deutsche Bundesbahn by 1984, but many went on to work for other operators in Greece, Switzerland, and Italy.

△ NSB Class Di3, 1955

Wheel arrangement	Co-Co
Transmission	electric
Engine	EMD 16-567-C
Total power output	1,750 hp (1,305 kW)
Top speed	65 mph (105 km/h)

The Swedish firm Nydqvist & Holm AB (NoHAB) built diesel locomotives under license for the major US diesel locomotive builder EMD, then owned by General Motors. As well as the Di3 locomotives, delivered in two types to Norwegian State Railways (Norges Statsbaner AS, or NSB), similar locomotives were supplied to Denmark and Hungary. The locomotives remain in service with freight operators in several European countries.

Diesel Switchers

While the big main-line diesel locomotives attracted the attention, using diesels in switching yards was just as transformational. While labor-intensive machines steam needed a team of operators and had to be kept "in steam" even when at rest, diesel switchers could be run by one person and simply turned off when not in use. Crew conditions were better too and, in many cases, so was the visibility from the cab. The advantages of the diesels were recognized even before World War II, and after the war their use became more and more widespread. Many of the 1950s designs had long working lives.

△ SNCF Class C61000, 1950

Wheel arrangement	0-6-0
Transmission	electric
Engine	Sulzer 6 LDA 22
Total power output	382 hp (285 kW)
Top speed	37 mph (60 km/h)

Ordered immediately after WWII in 1945, but not delivered until 1950-53, the 48 C61000 locomotives were used for switching in freight yards and for short-distance freight. Twelve of the locomotives were used with coupled powered "slave" units to double the power available for switching.

△ **DB VT11.5 (Class 601/602), 1957**

Wheel arrangement	B'2+2'2'+2'2'+2'2'+ 2'2'+2'2'+2'B
Transmission	hydraulic
Engine	2 x MTU engines
Total power output	2,060 hp (1,536 kW)
Top speed	100 mph (160 km/h)

The Class 601s were first-class-only diesel-powered train sets used for Trans-Europ Express services from 1957 to 1972, reaching Paris, Milan, Amsterdam, and Ostende. Some were rebuilt as Class 602 from 1970 with 2,145 hp (1,600 kW) gas turbines in place of the two diesel engines. The train sets were withdrawn from service in 1990.

△ **SNCF Class CC6500, 1957**

Wheel arrangement	Bo-Bo
Transmission	electric
Engine	2 x SACM MGO VSHR V12
Total power output	1,824 hp (1,360 kW)
Top speed	81 mph (130 km/h)

Because of their shape, these locomotives were nicknamed "Sous-marin" (submarines). Twenty were delivered to SNCF to replace steam locomotives in western France, where they worked until the 1980s; all were withdrawn by 1988. Tested widely when new, the design was also exported by the builder, Alsthom—37 to Algeria and 25 to Argentina.

◁ **DB VT98 (Class 798), 1955**

Wheel arrangement	single-car rail bus
Transmission	mechanical
Engine	2 x Büssing AG U10 engines
Total power output	295 hp (220 kW)
Top speed	56 mph (90 km/h)

These rail bus vehicles were introduced in West Germany from 1953 to 1962—initially the single-engined VT95 version, and then this more powerful two-engined VT98 version. In total, 913 powered and 1,217 unpowered trailer cars (of both types) replaced steam locomotives on many rural lines across West Germany.

◁ **PKP Class SM30, 1957**

Wheel arrangement	Bo-Bo
Transmission	electric
Engine	Wola V-300
Total power output	295 hp (220 kW)
Top speed	37 mph (60 km/h)

This was the first diesel-electric locomotive designed and built in Poland—its initial models used an engine originally designed for army tanks. Ultimately, 909 of the locomotives were built by Fablok in Chrzanów in southern Poland between 1956 and 1970, many for industrial users. Polish State Railways (PKP) received 302. Some are still in use in 2014.

△ **DR V15 (Class 101), 1959**

Wheel arrangement	0-4-0
Transmission	hydraulic
Engine	6 KVD 18 SRW
Total power output	148 hp (110 kW)
Top speed	22 mph (35 km/h)

The East German V15 (and later V18) diesel switchers were built in large numbers for both Deutsche Reichsbahn and industrial rail operators such as mines and steelworks. Built in Potsdam by VEB Lokomotivbau Karl Marx Babelsberg, many were also exported to other Eastern Bloc countries.

Great Journeys
The Blue Train

The Blue Train is one of the world's most luxurious trains. Styling itself as a "hotel on wheels," the train travels 994 miles (1,600 km) between Pretoria and Cape Town in South Africa and passes through scenery that ranges from lush vineyards to rugged semidesert.

THE PREDECESSORS OF *The Blue Train* came into service in the 1890s, picking up passengers from the Union-Castle liners docking in Cape Town and transporting them to the gold and diamond fields in the north. These early trains soon began catering to prospectors and wealthy travelers by offering more comfortable rail experiences. By 1923 the luxury Cape Town to Johannesburg trains were called the Union trains. The *Union Express* traveled from Cape Town to Johannesburg while the *Union Limited* made the return journey. By 1928 these trains offered facilities such as hot and cold water and heated cars, later acquiring dining saloons in 1933 and air-conditioning in 1939. The trains' distinctive, blue livery was introduced in 1936.

World War II caused train services to be suspended in 1942. They resumed in 1946, the same year that "those blue trains," as they were popularly known, adopted *The Blue Train* as their official designation. The trains have since been completely rebuilt twice, once in the 1970s and once in the 1990s.

Today the soundproofed, carpeted compartments all feature their own private bathrooms (luxury suites include full-size bathtubs). The train has underfloor heating, a restaurant car offering fine dining, two lounge cars, and an observation car (which converts to a conference car), as well as a 24-hour butler service and a laundry service.

Departing Cape Town
The Blue Train leaves the Cape to head north to Pretoria, flanked by the famous profile of Table Mountain.

**THE BLUE TRAIN
INSIGNIA**

The decor of each coach is unique, with birchwood paneling, marble finishes, and gold-plated fixtures throughout.

THE ROUTE TODAY

The Blue Train at one time traveled all the way to Victoria Falls in Zimbabwe, but this route has since been discontinued. Several others are now available but only as chartered services. The train's standard route from Cape Town to Pretoria runs through the Cape winelands and under the spectacular Hex River Mountains, where the train emerges from a series of tunnels into the arid region known as the Klein Karoo (Little Karoo). Here the train makes a stop at Matjiesfontein, a town that sprang up in 1884 around a refreshment station for passing trains, and remains preserved in its Victorian state. The service then continues on to Pretoria, passing through the semidesert landscape of the Great Karoo. On the return journey from Pretoria, passengers may disembark to visit the mining town of Kimberley, site of the diamond rush that began in the 1870s, before continuing on to Cape Town.

Lounging in luxury
There are two lavishly appointed lounges aboard the train, where passengers can expect five-star service. Once on board, food and drinks are all-inclusive.

KEY FACTS

DATES
1946 *The Blue Train* name is formally adopted
1970s, 1997 The train is refurbished

TRAINS
Train Set 1 Charter train; 14 cars accommodate 52 passengers
Train Set 2 Cape Town-Pretoria-Cape Town train; 19 cars accommodate 80 passengers
Locomotives 2 x 14E Class electric locomotives, dual current; 130 tons (120 metric tons)
Carriages 9 ft 5 in (2.9 m) wide—2 in (50 mm) wider than standard South African rolling stock. Thinner steel sides allow greater interior space
Speed 49 mph (80 km/h) with a maximum of 86 mph (138 km/h)
Weight Complete train 110 tons (100 metric tons)

JOURNEY
Cape Town to Pretoria (weekly) 994 miles (1,600 km) 27 hours
Cape Town to Durban (biannually, Sept & Nov) 473 miles (761 km) 21 hours. Also available for charter

RAILROAD
Gauge Cape Gauge 3 ft 6 in (1,067 mm)
Tunnels Four Hex River tunnels: twin tunnel 1,640 ft (500 m); single tunnels 3,609 ft (1,100 m), 3,937 ft (1,200 m), and 44,291 ft (13.5 km)
Bridges Orange River Station Bridge; Vaal River Crossing (Warrenton)
Highest point 5,751 ft (1,753 m), Johannesburg

Matjiesfontein [2]
This quaint and tiny museum town, little more than a single street, is now primarily a tourist destination.

The Hex River Valley [1]
The train passes through the vineyards of the valley and through four tunnels beneath the Hex (Witch) River Mountains, named for the girl who haunts them in local legend.

Great Karoo Desert

Table Mountain

Matjiesfontein

Paarl

Worcester

George

Cape Town

Kaaimans River [5]
The route crosses the Kaaimans estuary and passes through seven tunnels.

Victoria Falls

7 The Smoke that thunders
Victoria Falls, forming part of the border
of Zimbabwe and Zambia, is the world's
largest sheet of falling water. It was once
the spectacular conclusion to this now
discontinued route.

**Pretoria to Victoria
Falls (discontinued)**
991 miles (1,595 km);
two days, two nights.
The service ended in the
1990s due to political
unrest in Zimbabwe,
poorly maintained tracks,
and soaring rail tariffs.

Plumtree

Bulawayo

Z I M B A B W E

B O T S W A N A

Mahalapye

Kruger
National
Park

Gaborone

Hoedspruit

The Highveld
The highest point of the
rail journey is in the
Johannesburg area.

Bushveld break
Charter guests can stay
overnight at a private
game lodge.

Lobatse

Pretoria

**Pretoria to Hoedspruit
(charter only)**
Approx. 279 miles (450 km);
21 hours. Regular service was
discontinued in 2006.

Soweto

Johannesburg

S W A Z I L A N D

Klerksdorp

4 The Big Hole
The discovery of diamonds resulted
in what is reputedly the largest
human-made excavation to be created
solely with picks and shovels.

Valley of 1,000 Hills
Named for the rolling, green hills
beside the Umgeni River, this area,
visible from the train, was once the
battleground of the Zulu king, Shaka.

Kimberley

L E S O T H O

Pietermaritzburg

Durban

De Aar

S O U T H   A F R I C A

6 Durban beachfront
Durban's subtropical climate
and warm coastal waters
make it a year-round
vacation destination.

3 The Karoo
Much of the journey traverses the
semidesert of the Karoo, once a vast
inland sea, now a panoramic landscape of
scrub and "koppies" (low-topped hills).

**Cape Town to Port Elizabeth
(charter only, currently suspended)**
663 miles (1,067 km); two nights,
two days. Regular service was
discontinued in 2006.

Port Elizabeth

I N D I A N   O C E A N

| 0 | | 100 | | 200 miles |
| 0 | 100 | 200 | 300 km | |

N

KEY
● Start/Finish
● Main stations
||||| Main route
||||| Discontinued route
||||| Other routes

AN AFRICAN JOURNEY

The Blue Train travels through a range of terrain,
from the lush Cape to the arid Karoo. The
stopping points on the journey reflect South
Africa's colonial past and the source of the
country's wealth at the height of its powers.

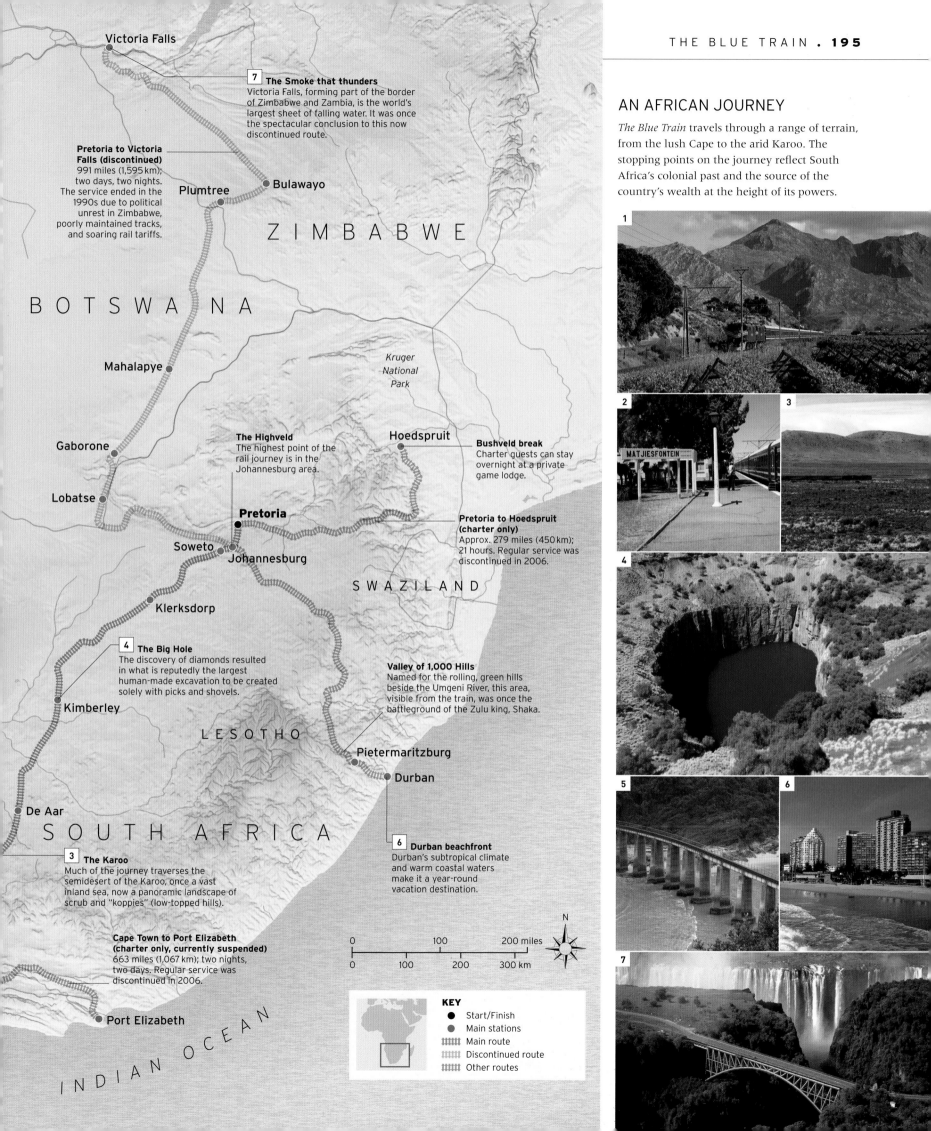

Electric Charge

In the early part of the 20th century, some European railroads had already started to use electric rather than steam locomotives on main lines in the Swiss and Austrian Alps—among the first to use this powerful new technology. Plans to expand electrified railroads were delayed almost everywhere in Europe by World War II, which led to the destruction of much railroad infrastructure. As postwar rebuilding got underway, most European countries turned to electrified railroads, and the 1950s saw new electric trains being widely introduced.

△ **BR Class 70 No. 20003, 1948**

Wheel arrangement Co-Co

Power supply 750 V DC third rail, overhead lines

Power rating 2,200 hp (1,641 kW)

Top speed 75 mph (121 km/h)

Following two similar locomotives (CC1/CC2) delivered to Southern Railway in 1941, No. 20003 was built at the Ashford Locomotive Works, Kent, in 1948 for British Railways. Like the earlier two, it was used until the late 1960s, mainly on the London to Brighton main line and other Sussex routes.

◁ **BLS Ae 4/4, 1944**

Wheel arrangement Bo-Bo

Power supply 15 kV AC, 16²/₃Hz, overhead lines

Power rating 3,950 hp (2,946 kW)

Top speed 78 mph (125 km/h)

Designed and built in Switzerland during WWII, the Ae 4/4 design was revolutionary, using a light steel body mounted on two-axle trucks. It produced nearly 4,000 hp (2984 kW), which was the equivalent of two or three steam engines. These design principles have been used for electric locomotives ever since.

▽ **SNCF Class BB9000, 1954**

Wheel arrangement Bo-Bo

Power supply 1,500 V DC, overhead lines

Power rating 4,000 hp (2,983 kW)

Top speed 206 mph (331 km/h)

This was one of two pairs of experimental express passenger engines using two-axle trucks that were delivered to the French state railroads in 1952-54: BB9003 and 9004 were built in France by Jeumont Schneider. On March 29, 1955, BB9004, along with CC7107, set a world record of 206 mph (331 km/h) for locomotives, which was not beaten until 2006.

BB-9004

S.N.C.F.

▷ BR Class EM1/ Class 76, 1954

Wheel arrangement Bo-Bo

Power supply 1,500 V DC, overhead lines

Power rating 1,868 hp (1,393 kW)

Top speed 65 mph (105 km/h)

Built for the electrification of the Manchester to Sheffield route via Woodhead, the first prototype was made for British Railways in 1940 but remained unused owing to WWII. It was tested in the Netherlands from 1947 to 1952, and was returned when the Woodhead line's electrification was completed.

◁ FS Class ETR, 1952

Wheel arrangement 7-car EMU

Power supply 3,000 kV DC, overhead lines

Power rating 3,485 hp (2,600 kW)

Top speed 124 mph (200 km/h)

Featuring a driving cab on the roof, a panoramic lounge at the front, and just 11 first-class seats, the "Settebello" ("Seven of Diamonds," named after an Italian card game) was the epitome of both high-speed and luxury travel. They were introduced by the Italian state railroads in the early 1950s. One "Settebello" train still exists.

▷ BR Class AL1/ Class 81, 1959

Wheel arrangement Bo-Bo

Power supply 25 kV AC, overhead lines

Power rating 3,200 hp (2,387 kW)

Top speed 100 mph (161 km/h)

This was the first production AC electric locomotive class built in the UK for the first British 25 kV AC main line electrification of the London to Birmingham/Manchester/Liverpool line. As BR Class 81, the locomotives remained in service until 1991.

▽ DB Class E41/141, 1956

Wheel arrangement Bo-Bo

Power supply 15 kV AC, 16²/₃ Hz, overhead lines

Power rating 3,218 hp (2,401 kW)

Top speed 75 mph (120 km/h)

Large-scale plans for electrification of West Germany's railroads during the 1950s led to large orders for several "universal" locomotive types built by consortiums comprising all the major German locomotive-building firms. The E41 was the "universal" design for light passenger and freight trains. In total, 451 were built between 1956 and 1971; all have now been withdrawn.

Postwar Steam

While railroads played a vital strategic role in Europe during World War II, the ravages of war, destruction of industry, and shortages of raw materials and fuel painted a bleak picture for the Continent's future. British railroads and workshops escaped the worst excesses of destruction, and with innovative locomotive designers, such as Oliver Bulleid and Robert Riddles, were introducing new types of successful austerity locomotives toward the end of the war. In contrast, on mainland Europe the national railroads were assisted in rebuilding their war-torn networks and rolling stock by deliveries of large numbers of powerful locomotives from US and Canadian manufacturers who were geared up for production through the Lend-Lease program and the 1948 Marshall Plan.

△ **SNCF 141R, 1945**

Wheel arrangement	2-8-2
Cylinders	2
Boiler pressure	225 psi (15.82 kg/sq cm)
Drive wheel diameter	65 in (1,650 mm)
Top speed	approx. 62 mph (100 km/h)

Powerful and economical to maintain, 1,323 Class 141R locomotives were built between 1945 and 1947 for the French state railroad (Société Nationale des Chemins de fer Français, or SNCF) by various builders in the US and Canada. Supplied under the Lend-Lease program to replace engines lost during WWII, around half were oil-burners. Many remained in service until the 1970s.

△ **Hunslet Austerity, 1944**

Wheel arrangement	0-6-0ST
Cylinders	2 (inside)
Boiler pressure	170 psi (11.95 kg/sq cm)
Drive wheel diameter	51 in (1,295 mm)
Top speed	approx. 35 mph (56 km/h)

Designed by the Hunslet Engine Co. of Leeds, these locomotives were chosen by the British War Department for use as its standard switching engine during WWII. Introduced in 1944, the earlier batches saw action in Europe and North Africa, as well as on military bases and ports across Great Britain.

▷ **SNCB 29, 1945**

Wheel arrangement	2-8-0
Cylinders	2
Boiler pressure	231 psi (16.24 kg/sq cm)
Drive wheel diameter	59 in (1,500 mm)
Top speed	approx. 60 mph (96 km/h)

After WWII these powerful mixed-traffic engines were built in Canada under the Lend-Lease program to help in the reopening of Belgium's ruined state railroad (Société Nationale des Chemins de fer Belges). Of the 180 built, one example, No. 29.013, has been preserved and is on display at the Belgian national railroad museum at Schaarbeek near Brussels.

◁ **SR Bulleid Light Pacific, 1945**

Wheel arrangement	4-6-2
Cylinders	3 (1 inside)
Boiler pressure	280 psi (19.68 kg/sq cm)
Drive wheel diameter	74 in (1,880 mm)
Top speed	approx. 80 mph (129 km/h)

Built under wartime conditions, Oliver Bulleid's "Battle of Britain" and "West Country" Class Light Pacific locomotives incorporated many cost-saving and innovative features. The 110 locomotives built for the Southern Railway and British Railways between 1945 and 1951 were renowned for their performance but suffered from high coal consumption. Sixty were subsequently rebuilt.

◁ GWR Modified Hall, 1944

Wheel arrangement 4-6-0

Cylinders 2

Boiler pressure 225 psi (15.82 kg sq cm)

Drive wheel diameter 70 in (1,778 mm)

Top speed approx. 75 mph (121 km/h)

With a large, three-row superheater to make up for the low-quality coal then available, these engines were a development by Frederick Hawksworth of Charles Collett's Hall Class. Between 1944 and 1950, a total of 71 were built at the Great Western Railway's Swindon Works.

◁ PKP Class Pt47, 1948

Wheel arrangement 2-8-2

Cylinders 2

Boiler pressure 213 psi (15 kg/sq cm)

Drive wheel diameter 72$\frac{3}{4}$in (1,850 mm)

Top speed approx. 68 mph (109 km/h)

Built by Fablok and Cegielski for the Polish state railroads (Polskie Koleje Panstwowe, or PKP) from 1948 to 1951, these engines achieved outstanding performances hauling heavy passenger trains over long distances.

△ SNCF 241P, 1948

Wheel arrangement 4-8-2

Cylinders 4 (2 high-pressure, 2 low-pressure)

Boiler pressure 284 psi (19.96 kg/sq cm)

Drive wheel diameter 78$\frac{3}{4}$in (2,000 mm)

Top speed 75 mph (121 km/h)

These powerful "Mountain"-type express passenger compound locomotives were built by Schneider for the French state railroad (SNCF) between 1948 and 1952. Designed to haul trains weighing 900 tons (800 metric tons) on the Paris to Marseilles main line, they were soon made obsolete by electrification.

◁ Andrew Barclay Industrial, 1949

Wheel arrangement 0-4-0ST

Cylinders 2

Boiler pressure 160 psi (11.25 kg/sq cm)

Drive wheel diameter 35$\frac{1}{2}$in (900 mm)

Top speed approx. 20 mph (32 km/h)

Scottish locomotive company Andrew Barclay built hundreds of these diminutive saddle tanks for use on private industrial railroads in Great Britain and abroad. Their short wheelbase enabled them to operate on the sharply curved lines at collieries, steel and gas works, and docks.

N&W J Class No. 611

No. 611 is the sole remaining example of the Norfolk & Western (N&W) Railway's mighty J Class 4-8-4s, built in Roanoke, VA, between 1941 and 1950. With its streamlined front end, large cylinders, and roller-bearings all around, the locomotive was built for running in excess of 100 mph (161 km/h) and regularly plied the N&W routes from Cincinnati to Norfolk and Portsmouth. Today, No. 611 is preserved at the Virginia Museum of Transportation.

THE STORY OF the N&W streamlined J Class 4-8-4s was in many respects defined by World War II. The engines were designed to haul N&W's prestigious, named express trains, such as the *Powhatan Arrow* and the *Pocahontas*, with the first five (Nos. 600–604) completed in 1941–42. However, their introduction came just as the US entered the war, and this was reflected in the second batch (Nos. 605–610), which was delivered in 1943. Due to wartime material shortages, these six locomotives were constructed without streamlining and light-weight rods.

The final, streamlined, batch (Nos. 611–613) did not appear until 1950, but their career was short-lived. By the late 1950s, N&W had begun experimenting with diesel locomotives, and steam was displaced by the end of the decade. Thanks in part to the efforts of the American railroad photographer O. Winston Link, No. 611 was spared the cutter's torch and donated to the Virginia Museum of Transportation, where it was returned to service in 1982.

FRONT VIEW

REAR VIEW

Fire Up 611

The builder's plate shows that 611 is a J Class completed in May 1950. Retired nine years later, 611 was eventually overhauled to pull excursion trains from 1982 until 1994. The "Fire Up 611" campaign is raising funds for a full restoration.

SPECIFICATIONS			
Class	J	**In-service period**	1950–59, 1982–94 (No. 611)
Wheel arrangement	4-8-4	**Cylinders**	2
Origin	United States	**Boiler pressure**	300 psi (21.09 kg/sq cm)
Designer/builder	Roanoke Shops	**Drive wheel diameter**	70 in (1,778 mm)
Number produced	14 J Class	**Top speed**	approx. 110 mph (177 km/h)

Tender carried by two 6-wheel trucks

"Tuscan red" stripe across full length of running board and tender

Firebox grate covers an area of 107 sq ft (10 sq m)

Roller bearings on all crank pins and axles for smoother running

Streamlined casing with bullet nose

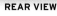

NORFOLK AND WESTERN   611

EXTERIOR

At 109 ft (33 m) in length, 16 ft (4.9 m) in height, and weighing over 436 tons (392 metric tons), No. 611 is an impressive locomotive and was the pride of the N&W Railway. Its purposeful lines are accentuated by the heavy appearance of its coupling and connecting rods.

1. Number plate (one on each side of headlight) 2. Chrome strips at front of casing 3. Headlight 4. Chrome marker lights 5. Front steps to running board 6. Control rod to throttle (regulator) 7. Sander valve 8. Handrail along running board 9. Air compressor under front side of engine 10. Lubrication system reservoir 11. Sander 12. Brake mechanism 13. Drive wheels and connecting rods 14. Injector 15. Cab window 16. Doors to tender coal bunker 17. Stoker screw inside tender

ENGINEER'S CAB

A locomotive of this size would be too much for a single fireman to manage using a shovel in the traditional style. Therefore, No. 611 was equipped with a mechanical stoker, which fed coal directly from the tender to the firebox by means of an Archimedes screw.

18. Cab interior **19.** Control levers for automatic grate shakers **20.** Control valves for stoker jets in firebox **21.** Gauge test valves and water level sight glass **22.** Open firebox door **23.** Staybolt detail inside firebox **24.** Circulators inside firebox **25.** Speedometer **26.** Brake control levers and handles **27.** Power reverse lever **28.** Electrical switches **29.** Throttle (regulator) quadrant **30.** Fireman's seat **31.** Footrest

World Steam's Last Stand

With seemingly unlimited supplies of cheap foreign oil, by the 1960s many European and North American railroads had replaced their steam engines with modern diesel-electric and electric engines, which were not only more efficient, powerful, and cleaner, but also required less maintenance between journeys. However, in other parts of the world where coal supplies were abundant and labor was cheap, steam continued to reign for a few more decades. In South Africa, the development of steam locomotive design reached its pinnacle in the 1980s with the "Red Devil." Ending in 2005, the awesome spectacle of QJ 2-10-2 double-headed freight trains running through the frozen wastes of inner Mongolia marked the final chapter of steam's 200-year reign.

△ N&W Class A, 1943

Wheel arrangement	2-6-6-4
Cylinders	4 (simple articulated)
Boiler pressure	300 psi (21.09 kg/sq cm)
Drive wheel diameter	70 in (1,778 mm)
Top speed	approx. 70 mph (113 km/h)

Built in the US at the Norfolk & Western Railroad's Roanoke Workshops, the Class A articulated fast freight engine was one of the most powerful in the world, remaining in service until 1959. Of the 43 built, one, No. 1218, is on display at the Virginia Museum of Transportation in Roanoke.

△ IR Class WP Pacific, 1947

Wheel arrangement	4-6-2
Cylinders	2
Boiler pressure	210 psi (14.8 kg/sq cm)
Drive wheel diameter	67 in (1,702 mm)
Top speed	approx. 68 mph (109 km/h)

Featuring a distinctive cone-shaped nose decorated with a silver star, 755 of the Class WP express passenger engines were built for the Indian broad-gauge railroads between 1947 and 1967. No. 7161 *Akbar* is preserved at the Rewari Steam Loco Shed, India.

◁ Soviet Class P36, 1949

Wheel arrangement	4-8-4
Cylinders	2
Boiler pressure	213 psi (15 kg/sq cm)
Drive wheel diameter	73 in (1,854 mm)
Top speed	approx. 78 mph (126 km/h)

Built between 1949 and 1956, the 251 Class P36 engines were the last Soviet standard class, first working on the Moscow to Leningrad line, until replaced by diesels. They later saw service in Eastern Siberia but were put into strategic storage from 1974 to the late 1980s.

▷ N&W J Class, 1950

Wheel arrangement	4-8-4
Cylinders	2
Boiler pressure	300 psi (21.09 kg/sq cm)
Drive wheel diameter	70 in (1,778 mm)
Top speed	approx. 70 mph (113 km/h)

A total of 14 J Class express passenger locomotives were built at the Norfolk & Western Railroad's Roanoke Workshops between 1941 and 1950. Fitted with futuristic streamlined casings, they were soon replaced by diesels and had all retired by 1959.

▷ UP Class 4000 "Big Boy," 1941

Wheel arrangement	4-8-8-4
Cylinders	4
Boiler pressure	300 psi (21.09 kg/sq cm)
Drive wheel diameter	68 in (1,730 mm)
Top speed	approx. 80 mph (129 km/h)

Twenty-five of these monster articulated locomotives were built by the American Locomotive Co. (ALCO) for the Union Pacific Railroad between 1941 and 1944. Nicknamed "Big Boys," they were designed to haul heavy freight trains unaided over the Wasatch Range between Wyoming and Utah. They were replaced by diesels in 1959. Eight have been preserved, of which No. 4014 is being restored to working order.

△ IR Class YG, 1949

Wheel arrangement	2-8-2
Cylinders	2
Boiler pressure	210 psi (14.78 kg/sq cm)
Drive wheel diameter	48 in (1,220 mm)
Top speed	approx. 50 mph (80 km/h)

The Class YG was the standard freight locomotive on the Indian Railways 3-ft 3-in- (1-m-) gauge system. Around 1,000 were built by various manufacturers in India and overseas between 1949 and 1972. Three, including *Sindh* (seen here), are preserved in working order at Rewari Steam Loco Shed southwest of Delhi.

TALKING POINT

Cutting-edge Steam

Apart from the Class 25 condensing engines, South African Railways also took delivery of 50 Class 25NC (non-condensing). Of these, No. 3450 was modified in 1981 at the SAR's Salt River Workshops in Cape Town as the prototype Class 26. Nicknamed the "Red Devil" because of its livery, tests demonstrated vastly increased power and savings, diesel and electric traction had virtually replaced steam by the early 1980s.

The "Red Devil" This unique engine is seen here leaving Krankuil with a South African rail tour in 1990. It last ran in 2003 and is now preserved in Cape Town.

▷ SAR Class 25C, 1953

Wheel arrangement	4-8-4
Cylinders	2
Boiler pressure	225 psi (15.81 kg/sq cm)
Drive wheel diameter	60 in (1,524 mm)
Top speed	approx. 70 mph (113 km/h)

A total of 90 Class 25C locomotives were built for the 3-ft-6-in- (1.06-m-) gauge South African Railways. The engines were originally equipped with an enormous condensing tender so that they could operate across the arid Karoo Desert. Most were later converted to a non-condensing Class 25NC between 1973 and 1980.

△ IR Class WL, 1955

Wheel arrangement	4-6-2
Cylinders	2
Boiler pressure	210 psi (14.76 kg/sq cm)
Drive wheel diameter	67 in (1,702 mm)
Top speed	approx. 60 mph (96 km/h)

Featuring a light axle load for work on branch lines, these broad-gauge steam engines were built for the Indian Railways in two batches: the first 10 by Vulcan Foundry (UK) and 94 at the Chittaranjan Locomotive Works (India). No. 15005 *Sher-e-Punjab* is preserved at Rewari.

▷ China Railways CS Class QJ, 1956

Wheel arrangement	2-10-2
Cylinders	2
Boiler pressure	213 psi (15 kg/sq cm)
Drive wheel diameter	59 in (1,500 mm)
Top speed	approx. 50 mph (80 km/h)

One of the most prolific classes constructed in China was the Class QJ heavy freight engine, of which at least 4,700 were built between 1956 and 1988. Their service on the Jitong Railway in Inner Mongolia (China) ended in 2005, though some ran on industrial railroads until 2010.

Class WP No. 7161

Manufactured in the US by the Baldwin Locomotive Works, the first 16 Class WP steam locomotives (W for 5-ft 6-in broad gauge and P for passenger) arrived in India in 1947. Chittaranjan Locomotive Works of West Bengal built No. 7161 in 1965 to run on the Northeast Frontier Railway. Now based at Rewari Steam Loco Shed, where it was named *Akbar* after the legendary Mughal emperor, it is the only working locomotive of its class.

WHEN THE CLASS WP of sleek, bullet-nosed, mainline steam locomotives was first introduced, it set the standard on Indian railroads and became the mainstay of broad-gauge passenger operations for the rest of the 20th century. Known for free steaming, high fuel economy, and superior riding characteristics—and without the tail wag of the earlier X classes—its arrival marked the change of broad gauge coding from X to W.

Initially imported until 1959, the Class WP was manufactured in India between 1963 and 1967 at the Chittaranjan Locomotive Works, where 259 engines were built. Requiring a crew of three—a driver and two firemen—these locomotives hauled most of the prestigious passenger trains on the Indian rail system for the next 25 years. They established a sound reputation during their time in service, with their good performance earning them the title "Pride of the Fleet."

FRONT VIEW

REAR VIEW

Heritage shed
Converted to a heritage museum by Indian Railways in 2002, the Rewari Steam Loco Shed houses some of India's last surviving steam locomotives.

SPECIFICATIONS			
Class	WP	**In-service period**	1965–96 (No. 7161)
Wheel arrangement	4-6-2	**Cylinders**	2
Origin	India	**Boiler pressure**	210 psi (14.78 kg/sq cm)
Designer/builder	Chittaranjan Locomotive Works	**Drive wheel diameter**	67 in (1,702 mm)
Number produced	755 (259 in India) Class WP	**Top speed**	68 mph (109 km/h)

Tender could carry 18 tons of coal and 6,500 gallons (24,605 liters) of water

Cab accommodates a crew of three

Air brake pipes outside locomotive frame

Metal chains along the length of the running board

Stack is topped by a decorative crown

Bullet nose mounted on smoke box

India's star locomotive
The bullet nose bears a silver star and is the most distinctive feature of the locomotive. A nameplate with the name *Akbar* sits centrally below the nose.

EXTERIOR

With its distinctive bullet nose, a crown on top of the stack, a 4-6-2 wheel arrangement, and a side profile that includes chain-decorated footboards along the length of the boiler, No. 7161 is regarded as one of the most majestic locomotives that has ever run on Indian Railways. These features have proved popular with railroad enthusiasts and tourists, and the locomotive currently hauls a main-line tourist train.

1. Hand-painted name on plaque at front **2.** Brass crown decorating top of smokestack **3.** Headlight in the center of a metal star **4.** Pilot light lamp, one positioned on either side of engine **5.** Cattle guard **6.** Steam chest valves **7.** Steam chest **8.** Driving wheels, with balance weight and connecting rod **9.** Big end motion **10.** Rear carrying wheel **11.** Steps leading to cab interior **12.** Entrance into engineer's cab with wooden-slatted windows **13.** Light at back of tender **14.** Engine number **15.** Ladder at back of tender **16.** Rear buffer

CAB INTERIOR

The cab is spacious enough to house the engineer and two firemen. The extra space also allows the firemen to stoke coal using shovels that are larger than those used in earlier locomotives. The red-painted handles for operating the locomotive and the monitoring gauges are positioned for ease of use.

17. Interior of engineer's cab **18.** Lubricator box **19.** From left to right: injector steam cock handle, dynamo cock, main cock, vacuum steam cock, and injector steam cock handle **20.** Steam pressure gauge **21.** Reverser wheel **22.** Firehole door **23.** Rocking grate **24.** Front of tender

Europe's Last Gasp

With diesel and electric traction rapidly gaining favor, the 1950s saw the last steam locomotives built for Europe's national railroads. In West Germany, the last one to be built for Deutsche Bundesbahn, No. 23.105, rolled off the production line in 1959. Across the English Channel, Robert Riddles had designed 12 new classes of standard locomotives for the nationalized British Railways. Sadly, many of these fine engines had extremely short working lives owing to the hurried implementation of the ill-conceived Modernization Plan. Despite this, privately owned British locomotive manufacturers such as Beyer Peacock & Co. of Manchester and Hunslet of Leeds continued to export steam locomotives; the last engine was built by Hunslet in 1971.

◁ DB Class 23, 1950

Wheel arrangement	2-6-2
Cylinders	2
Boiler pressure	232 psi (16.3 kg/sq cm)
Drive wheel diameter	69 in (1,750 mm)
Top speed	68 mph (110 km/h)

This engine was designed to replace the Prussian Class P8 passenger locomotives on the West German Deutsche Bundesbahn. The 105 Class 23s were built between 1950 and 1959. No. 23.105 was the last steam locomotive built for DB. The final examples were retired in 1976, and eight have been preserved.

△ *Bonnie Prince Charlie*, 1951

Wheel arrangement	0-4-0ST
Cylinders	2
Boiler pressure	160 psi (11.25 kg/sq cm)
Drive wheel diameter	24 in (610 mm)
Top speed	20 mph (32 km/h)

Built by Robert Stephenson & Hawthorns in 1951, *Bonnie Prince Charlie* originally worked as a gasworks switcher at Hamworthy Quay, Dorset, England. It was bought by the Salisbury Steam Trust in 1969 and has since been restored at Didcot Railway Centre.

▷ BR Class 4MT, 1951

Wheel arrangement	2-6-4T
Cylinders	2
Boiler pressure	225 psi (15.82 kg/sq cm)
Drive wheel diameter	68 in (1,730 mm)
Top speed	70 mph (113 km/h)

Robert Riddles' Class 4MT tank locomotive was the largest of four standard tank designs built by British Railways. Used primarily on suburban commuter services, a total of 155 were built between 1951 and 1956, but they were soon displaced by electrification.

▷ BR Class 9F, 1954

Wheel arrangement	2-10-0
Cylinders	2
Boiler pressure	250 psi (17.57 kg/sq cm)
Drive wheel diameter	60 in (1,524 mm)
Top speed	90 mph (145 km/h)

The Class 9F was the standard heavy freight locomotive built by British Railways between 1954 and 1960. A total of 251 were built, with No. 92220 *Evening Star* being the last steam engine built for BR. Although designed to haul freight, they were occasionally used on express passenger duties. All were retired by 1968, and nine have been preserved.

▷ **BR Class 7 Britannia, 1951**

Wheel arrangement	4-6-2
Cylinders	2
Boiler pressure	250 psi (17.57 kg/sq cm)
Drive wheel diameter	74 in (1,880 mm)
Top speed	90 mph (145 km/h)

A total of 55 Class 7 Britannia engines, designed by Robert Riddles, were built at British Railways' Crewe Works between 1951 and 1954. After hauling expresses across the BR network, they were relegated to more humble duties. One lasted until the end of BR mainline steam in 1968.

△ **DR Class 65.10, 1954**

Wheel arrangement	2-8-4T
Cylinders	2
Boiler pressure	232 psi (16.3 kg/sq cm)
Drive wheel diameter	63 in (1,600 mm)
Top speed	56 mph (90 km/h)

The powerful Class 65.10 tank locomotives were built to haul double-deck and push-pull commuter trains on the Deutsche Reichsbahn in East Germany. All 88 built had retired by 1977, but three have been preserved.

◁ **DR Class 99.23-24, 1954**

Wheel arrangement	2-10-2T
Cylinders	2
Boiler pressure	203 psi (14.27 kg/sq cm)
Drive wheel diameter	39½ in (1,000 mm)
Top speed	25 mph (40 km/h)

Seventeen of these massive 3-ft-3-in- (1-m-) gauge tank locomotives were built for Deutsche Reichsbahn between 1954 and 1956. They all still survive on the highly scenic railroads in the Harz Mountains, with nine currently in working order.

△ **Beyer-Garratt Class NG G16, 1958**

Wheel arrangement	2-6-2+2-6-2
Cylinders	4
Boiler pressure	180 psi (12.65 kg/sq cm)
Drive wheel diameter	33 in (840 mm)
Top speed	40 mph (64 km/h)

Several European manufacturers built 34 of these engines from 1937 to 1968 for the 2-ft- (0.60-m-) gauge lines of South African Railways. No.138, built by Beyer Peacock & Co., now hauls trains on the Welsh Highland Railway.

Beyer-Garratt No.138

The NG G16 class of Beyer-Garratts has achieved international prominence serving the Welsh Highland Railway, but these locomotives were originally built for mining companies in southern Africa. Used mainly for freight in Africa, in Wales these engines have a new life pulling passenger cars in Snowdonia National Park, showing off their haulage capacity and articulation on the steep gradients and sharp curves of the mountainous landscape.

BEYER-GARRATT NO.138 is one of the last batch of Garratt locomotives built by Beyer, Peacock & Company Ltd. in Manchester, England. The Tsumeb Corporation of South West Africa (now Namibia) ordered seven locomotives of this type to haul minerals from its mines in the Otavi mountains. However, the regauging of the 256-mile- (412-km-) long Otavi Railway to 3 ft 6 in (106 cm) before the locomotives arrived resulted in their sale to South African Railways for use in Natal, on the east coast.

Allocated to the 122-km- (76-mile-) long Port Shepstone–Harding line, No.138 was one of the assets transferred when the railroad was privatized in 1986. The locomotive was withdrawn from service in 1991. However, after being selected by the Ffestiniog Railway for use on the Welsh Highland Railway (WHR) in 1993, it was overhauled at Port Shepstone and delivered to Wales. It began running on the WHR in green livery in October 1997, but was painted red in 2010.

FRONT VIEW

REAR VIEW

WHR

Preserving history
At 25 miles (40 km), the Welsh Highland Railway is the longest heritage railroad in Great Britain. It stopped running before World War II, but a restoration project was completed in 2011.

SPECIFICATIONS				
Class	NG G16	In-service period	1958–91 and 1997 to present (No.138)
Wheel arrangement	2-6-2+2-6-2	Cylinders	4
Origin	UK	Boiler pressure	180 psi (12.65 kg/sq cm)
Designer/builder	Beyer, Peacock & Co. Ltd.	Drive wheel diameter	33 in (840 mm)
Number produced	34 Class NG G16	Top speed	approx. 40 mph (64 km/h)

Water tank holds up to 1,590 gallons (6,023 liters)

Chimney rises 10 ft 4 in (315 cm) above the rails

Boiler is slung on a cradle between the two engines

Stainless steel bands hold boiler jacket in place

Number plate is written in English and Welsh

Coal bunker sits on top of rear engine, and replaces separate tender

A touch of style
The curves and angles of No.18.201 gave the locomotive a stylishly modern look. The smokebox door was a distinctive conical shape, while the locomotive's thin but efficient *Giesl* ejector exhaust was hidden inside a much larger stack shroud.

EXTERIOR

Although unique and instantly recognizable, No. 18.201 shares design elements with other German steam locomotives, as well as high-speed engines from elsewhere. The look is dominated by the green semi-streamlined casing and the large 91-in (2,311-mm) drive wheels, which that lend additional speed to the engine. Some details, such as the front headlights, are non-standard add-ons, while others have come from the Deutsche Reichsbahn stores.

1. Number plate on side of cab **2.** Front headlight **3.** Coupling hook **4.** Front buffer **5.** Front steps **6.** Steam-powered electrical generator **7.** Shut-off valve **8.** Whistle **9.** Lagged pipework **10.** Valve gear **11.** Small End **12.** Truck wheel **13.** Inside the Big End **14.** Lead drive wheel **15.** Sand pipe **16.** Brake assembly **17.** Air pump assembly **18.** Steps to cab at front of tender **19.** "Indusi" magnet **20.** Detail of tender bogie **21.** Headlight on rear of tender **22.** Oil filler

CAB INTERIOR

In comparison with coal-fired locomotives, the oil-fired No. 18.201 has different controls for the fireman to regulate the fire, as well as dials to monitor it. The insulated firebox door stays shut while the engine is operating, which makes the onerous task of shoveling unnecessary. The engineer sits on the right, where all the main driving controls are within easy reach.

23. Overview of cab controls **24.** Lubricator **25.** Light switches **26.** Pressure gauge **27.** Sanding controls **28.** Reverse/cut-off indicator **29.** Fireman's seat **30.** Reverser **31.** Firebox door **32.** Interior of firebox **33.** Area at front of tender

Technology in Transition

This was a period of large-scale changes for railroads around the world. Car ownership and the impact of new highways led to the closure of less-used rail routes, particularly in Western Europe, although branch lines still thrived in Eastern Europe. Commodities such as coal and iron ore continued to be carried by the railroads. Much local freight shipping switched to trucks, but the use of intermodal containers to carry long-distance freight by rail continued to grow. In addition, many European cities were expanding existing metro systems or building new ones.

△ BR D9500 (Class 14), 1964

Wheel arrangement	0-6-0
Transmission	hydraulic
Engine	Paxman 6YJXL
Total power output	650 hp (485 kW)
Top speed	40 mph (64 km/h)

The 56 locomotives in this class, all built at the British Railway Works in Swindon during 1964, were delivered just as the local freight traffic they were designed for was rapidly disappearing from the UK rail network. As a result, many were withdrawn within three years. Most went on to have longer careers with industrial rail operators in the UK and Europe.

◁ Soviet Class VL10, 1963

Wheel arrangement	Bo-Bo+Bo-Bo
Power supply	3,000 V DC, overhead lines
Power rating	6,166 hp (4,600 kW)
Top speed	62 mph (100 km/h)

Built in Tbilisi (now Georgia), the VL10, eight-axle, twin-unit electric was the first modern DC electric locomotive built for Soviet Railways. It shared both external design and many components with the VL80 25 kV AC electric design, also introduced in 1963. Thousands of both classes were built until production ended in the 1980s.

△ DR VT2.09 (Class 171/172), 1962

Wheel arrangement	2-axle rail bus
Transmission	mechanical/hydro-mechanical
Engine	6 KVD 18 HRW
Total power output	180 hp (134 kW)
Top speed	56 mph (90 km/h)

Designed for East Germany's rural branch lines, this train was nicknamed "Ferkeltaxi" (piglet taxi) because farmers sometimes brought piglets along as luggage. An early prototype built in 1957 was followed by orders for production trains, delivered from 1962 to 1969. In 2004 they were withdrawn from regular use in Germany.

Track Maintenance

Motorized draisines replaced or supplemented daily track inspections carried out on foot from the 1960s, allowing tools and equipment to be carried to work sites quickly. During the 1960s, ultrasonic testing of rails by test vehicles with special equipment became more common in both US and Europe, and regular test trains operated, often at night, to monitor track condition.

Room for two In East Germany, the two-axle draisine could carry two people and their tools to repair minor faults.

▷ DR V60 D (Class 105), 1961

Wheel arrangement	0-8-0
Transmission	hydraulic
Engine	12 KVD 18/21
Total power output	650 hp (485 kW)
Top speed	37 mph (60 km/h)

The powerful V60 was designed to replace the Deutsche Reichsbahn steam locomotives for switching and short freight trains. The engines, enhanced by advances made on the WWII V36 diesels used by the German military, unusually had four axles with the wheels connected by external coupling rods. They were built for the DR and other state railroads plus heavy industry in Eastern Bloc countries.

TECHNOLOGY

Battery Locomotives

In many European countries, battery-powered engines were used to move locomotives around maintenance depots. Using the battery engines allowed electric locomotives to be transferred to maintenance areas without (hazardous) overhead power lines for traction current, and was quicker and cheaper than starting a diesel to move it a few hundred yards. Battery engines continue to be used in this way today.

Akkuschleppfahrzeuge (ASF) Over 500 ASFs (meaning "battery switching vehicle") were built in East Germany from 1966 to 1990. Used by DR and industrial operators, some are still working.

△ **LT Victoria Line, 1969**

Wheel arrangement 4-car units, always operated as pairs	
Power supply 630 V DC third and fourth rail system	
Power rating 1,137 hp (848 kW)	
Top speed 25 mph (40 km/h)	

The Victoria Line was the first completely new Tube line in London for 60 years when it opened in 1969. The new trains bought by London Transport were equipped with Automatic Train Operation (ATO) equipment—the train drove itself and the "driver" would normally only open and close doors at stations.

△ **Preston Docks Sentinel, 1968**

Wheel arrangement A-A	
Transmission hydraulic	
Engine Rolls-Royce C8SFL	
Total power output 325 hp (242 kW)	
Top speed 18 mph (29 km/h)	

The Sentinel locomotives were designed to replace steam engines at major industrial sites that operated their own railroads. Innovative and easy to use, they had a central driving position in a full-width cab and safe places for switching staff to travel on the outside of the engines. Several are preserved on British heritage railroads.

▽ **DR V300 (Class 132), 1973**

Wheel arrangement Co-Co	
Transmission electric	
Engine Kolomna 5D49	
Total power output 3,000 hp (2,238 kW)	
Top speed 74 mph (120 km/h)	

Based on the Soviet TE109 design and built at Voroshilovgrad (now Luhansk, Ukraine), the most numerous of the DR V300 locomotives was Class 132, with 709 locomotives. While most have been withdrawn, some remain in service with several German freight operators today.

Great Journeys
Indian Pacific

The first direct passenger rail service to cross the continent of Australia from the east coast to the west, the *Indian Pacific* finally linked Sydney on the Pacific Ocean to Perth on the Indian Ocean on February 23, 1970.

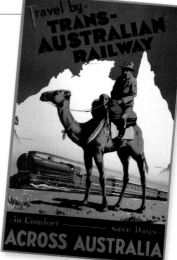

AUSTRALIA'S FIRST UNBROKEN transcontinental railroad was made possible only by standardizing the random mixture of broad-, standard-, and narrow-gauge lines that had been built in the 19th and early 20th centuries.

The New South Wales government opened the state's first standard-gauge railroad in 1855, linking Sydney on the east coast to nearby Granville. This track was gradually extended over the Blue Mountains via a series of steeply graded zigzags, reaching Orange—200 miles (322 km) from Sydney—in 1877. From Orange, the standard-gauge Broken Hill Line opened westward in stages, between 1885 and 1927, across sparsely populated, arid lands to the mining town of Broken Hill. Westward from Broken Hill, the 3-ft 6-in (1.06-m) gauge Silverton Tramway, opened in 1888, reached as far as

Indian Pacific stops in Broken Hill
A 4,000-hp (2,934-kW) NR Class diesel-electric locomotive pulling the *Indian Pacific* stops at Broken Hill. The town is at the center of the world's largest silver, lead, and zinc ore deposits.

Saving days by train
A vintage travel poster by Australian artist James Northfield publicizes the advantages of the newly built Trans-Australian section of the railway.

Cockburn. Here the railroad met the South Australian Railways' line of the same gauge from Port Pirie, part of the Adelaide to Port Augusta line.

From the west coast, a 3-ft 6-in (1.06-m) gauge line had already linked Perth to the gold-mining town of Kalgoorlie by 1897. Between Kalgoorlie and Port Augusta remained a 1,000-mile (1,609-km) gap across South Australia through a region that was a virtually uninhabited and waterless desert.

In 1901 the newly formed Commonwealth of Australia's government proposed a railroad to link isolated Western Australia with the rest of the country. Opened throughout in 1917, the 1,052-mile (1,693-km) Trans-Australian Railway across the aptly named Nullarbor ("no tree") Plain was built to the standard gauge of 4 ft 8½ in (1.435 m), but met narrow-gauge lines at either end.

KEY
- ● Start/Finish
- ● Main stations
- ⅢⅢⅢ Earlier lines
- ⅢⅢⅢ Trans-Australian Railway

AUSTRALIA

SOUTH AUSTRALIA

WESTERN AUSTRALIA

World's longest straight stretch of track This section is 297 miles (478 km).

Perth to Kalgoorlie
Originally built to the 3-ft 6-in (1.06-mm) gauge, this line was completed in 1897.

Kalgoorlie

Rawlinna

Loongana

Nullarbor Plain

Ooldea

Tarcoola

1 Camp train
Laborers building the Trans-Australian lived in mobile accommodations on rails, to avoid constantly breaking camp as the track advanced.

Meeting of two teams
The two halves of the Trans-Australian line met at Ooldea on October 17, 1917.

Perth

6 Perth
Until the opening of the Trans-Australian Railway in 1917, Perth could be reached from the east only by a sea voyage across the Great Australian Bight.

Nullarbor Plain 5
The Trans-Australian line's construction faced a huge challenge in passing through an almost waterless region.

Port Augusta to Port Pirie
Converted from 3-ft 6-in (1.06-mm) gauge to standard gauge in 1937.

Port Augusta

Port Pirie

Adelaide

GREAT AUSTRALIAN BIGHT

Narrow gauge The 3-ft 6-in (1.06-m) gauge line from Port Pirie to Adelaide was converted to standard gauge in 1982.

0   100   200   300 miles
0   100   200   300   400 km

N

No natural water sources existed on this stretch of the line, so steam-hauled trains had to carry their own supplies, which occupied over half the train's load. Diesels took over in 1951.

A unified standard-gauge railroad across the continent was created in stages: the line from Port Augusta to Port Pirie was converted in 1937, and the 374-mile (602-km) line from Perth to Kalgoorlie was converted in 1969. The track between Port Pirie and Broken Hill was rebuilt as standard gauge by 1970, and the *Indian Pacific* made its first run from Sydney. Now a luxury train, it completes the four-day journey twice a week, stopping off at the historic Broken Hill and offering an experience of remote Australian terrain.

In 1982 the *Indian Pacific* also began to call at Adelaide after the line south of Port Pirie was converted to standard gauge, extending the distance traveled by the train to 2,704 miles (4,352 km).

Across the Nullarbor
Double-headed by NR Class diesels, the *Indian Pacific* heads out across the arid Nullarbor Plain on the world's longest straight stretch of track.

KEY FACTS

DATES
1917 Standard-gauge Trans-Australian Railway completed, meeting existing narrow-gauge lines in east and west.
1970 Continuous standard-gauge railroad between Sydney and Perth completed. *Indian Pacific* inaugural run on February 23.

TRAIN
First locomotives Commonwealth Railways CL Class 3,000 hp (2,238 kW) Co-Co diesel-electrics built 1970–72
Current locomotives NR class 4,000 hp (2,984 kW) Co-Co diesel-electrics built 1996–98
Cars Up to 25 75-ft (23-m) air-conditioned stainless steel cars, including sleeper cars, restaurant car, power van, luggage van, and Motorail cars carrying passengers' cars. Three classes: Platinum, Gold, and Red service.

JOURNEY
Original journey: Sydney–Perth 2,461 miles (3,961 km); 75 hours
Sydney–Perth (via Adelaide) 2,704 miles (4,352 km), 65 hours; 4 days, 3 nights

RAILROAD
Gauge Standard gauge 4 ft 8 ½ in (1.435 m)
Longest straight stretch The world's longest section of straight track, 297 miles (478 km)
Highest point Bell train station in the Blue Mountains: 3,507 ft (1,069 m)

ACROSS A CONTINENT

On its 65-hour journey across New South Wales, South Australia, and Western Australia, the *Indian Pacific* crosses three time zones, which were introduced in the 1890s. Perth is two hours behind Sydney in summer, three in winter.

QUEENSLAND

4 Mannahill Station
Located along the *Indian Pacific* route, Mannahill is one of the easternmost settlements in South Australia. It has only 66 inhabitants.

NEW SOUTH WALES

Broken Hill

Blue Mountains The rail routes originally crossed the mountains on steeply graded zigzags, which were bypassed in 1910.

Cockburn Ivanhoe

Condobolin

Granville

3 Sydney
Australia's premier east coast city is famous for its iconic Opera House and Harbour Bridge.

Orange

AUSTRALIAN CAPITAL TERRITORY

Sydney

PACIFIC OCEAN

Broken Hill to Port Pirie
Converted from 3-ft 6-in (1.06-m) gauge to standard gauge in 1970.

VICTORIA

2 Inaugural journey
The *Indian Pacific*, the first train to cross the entire Australian continent, left Sydney on February 23, 1970.

Traveling in Style

In the 1960s and 70s, railroads around the world invested in large numbers of new passenger cars. The investment was partly driven by the need to offer higher speed and more comfort on intercity routes, and in other cases simply to replace older equipment. Steel became the dominant material for coach bodies, replacing wooden-framed, steam-age vehicles in many cases. Increasing numbers of new multiple-unit trains, both diesel and electric, were built in many countries to replace conventional trains using locomotives and coaches.

◁ **Cravens Stock, 1963**

Type second class open coach
Capacity 64 passengers
Construction steel
Railway CIÉ (Irish Railways)

Fifty-eight of these coaches were assembled in Irish Railways's Inchicore Works in Dublin between 1963 and 1967, using kits provided by Cravens in Sheffield, England. The coaches were equipped with steam heating and vacuum brakes, and were used for express trains in the 1960s. Several coaches have been preserved.

△ **Talgo III, 1964**

Type articulated express passenger car
Capacity 21 passengers
Construction stainless steel
Railway RENFE (Spanish state railroads)

In the 1950s the Spanish Talgo company pioneered articulated trains of semi-permanently coupled short cars utilizing single-axle wheel sets. The Talgo III was the third version of the train and the first to be used internationally. Some had variable-gauge axles, which permitted operation from Spain into France.

△ **Penn Central/Amtrak Metroliner, 1969**

Type snack bar car (powered)
Wheel arrangement 2-car EMU
Power supply 11 kV AC 25 Hz, 11 kV AC 60 Hz, and 25 kV AC 60 Hz, overhead lines
Power rating 1,020 hp (761 kW)
Top speed 125 mph (200 km/h)

Budd built 61 Metroliner EMU cars for Penn Central Transportation in 1969 in collaboration with other manufacturers and the US government. The cars were inherited by Amtrak in 1971. Designed for use at 150 mph (241 km/h), the Metroliners never operated that fast, and most were withdrawn by Amtrak in the 1980s.

△ *Reko-Wagen*, 1967

Type second-class open coach
Capacity 64 passengers
Construction steel
Railway Deutsche Reichsbahn

The Deutsche Reichsbahn introduced *Reko-Wagen* (reconstructed coaches) in the 1950s and 60s—the reconstruction referred to their rebuild from older designs. Initially, short three-axle coaches were built, but in 1967, 61-ft- (18.7-m-) long truck coaches appeared.

△ **Eurofima, 1973**

Type first and second class open
Capacity 54 (first); 66 (second)
Construction steel
Railway SBB (Swiss Railways, and others)

In the mid-1970s several Western European railroads jointly ordered 500 new daytime coaches to a standard design following tests with 10 prototypes. They were funded via Eurofima, a nonprofit rail financing organization based in Switzerland. In total, 500 coaches were built for six different operators.

▽ Mark IIIB First Open, 1975

Type first-class Pullman coach

Capacity 48 passengers

Construction steel

Railway British Rail

The first 125 mph (201 km/h) Mark III coaches appeared in 1975 and incorporated steel integral monocoque construction, giving them great body strength. The BR High Speed Train (HST) used Mark III coaches and others were built for use with electric locomotives at up to 110 mph (177 km/h).

△ Mark III sleeper, 1979

Type sleeping coach

Capacity 26 berths in 13 compartments

Construction steel

Railway British Rail

In 1976 British Rail ordered a new prototype sleeper to replace its older cars, but this was canceled after a fatal fire on Mark I sleepers on an overnight train in Taunton, Somerset, in 1978. BR decided to build a new version that incorporated safety systems into all sleepers; 236 were ordered in 1979.

▷ Amtrak Superliner, 1978

Type double-deck long distance

Capacity up to 74, fewer for sleepers

Construction stainless steel

Railway Amtrak

Based on cars originally built in 1956 for the Atchison, Topeka & Santa Fe Railway and inherited by Amtrak in 1971, the Superliner long-distance cars were built from 1978. Nearly 500 were made over the next 20 years in multiple configurations (sleepers, seating cars, diners, and observation cars).

Danger
Overhead
live wires

Class	92	A
Weight tonnes		126
Brake force tonnes		53
ETH index		:
ac		180
dc		108
RA		7
Max speed km/h		140
Braked Weight		:
Goods tonnes		65
Passenger tonnes		94

**Cab 1
Side B**

1980-1999
CHANGING TRACKS

CHANGING TRACKS

The high-speed railroad spread internationally as more countries built dedicated networks replicating the Japanese invention. In Europe, France's Train à Grande Vitesse (TGV) was launched in 1981 with a line running from Paris to Lyon, and a decade later Germany saw the InterCityExpress (ICE) make its public debut. In the UK, however, the emphasis lay on modernizing the existing system, rather than building new lines.

As the renaissance in light rail continued, new tram (streetcar) systems opened in some places. Karlsruhe in southwest Germany introduced a new concept: the "tram-train," a vehicle capable of running both on the streets and on local train tracks. Yet while rail technology improved, there was also a desire for "golden age" travel inspired by the past, which was realized with the launch of classic luxury trains such as Europe's *Venice Simplon-Orient Express* and India's *Palace on Wheels*.

The end of the Cold War ushered in changes to Europe's railroads, not least in Germany. Following the country's unification in 1990, lines that ran across the former border were reopened and new ones were built, and the former East and West German systems were eventually merged as the Deutsche Bahn.

However, the restructuring was much more radical in the UK, after the British parliament voted for privatization in 1993. In the years that followed, the state-owned British Rail was dismantled; new companies took over different lines and implemented their own plans for development, and reintroducing variety to train services.

In 1994, rail celebrated yet another engineering marvel with the opening of the Channel Tunnel, which connected France and England for the first time. Running under the Dover Strait, the launch of the tunnel was the realization of a dream dating back to the 19th century.

"So speed yes, but let there be money in it"

GERARD FIENNES,
FIENNES ON RAILS, 1986

△ **Across the Channel**
On November 14, 1994 Eurostar services began between London Waterloo International, Paris Gare du Nord, and Brussels-South.

Key events

▷ **1981** High-speed rail services come to Europe when France launches the TGV—the Train à Grande Vitesse. TGV raises the world speed record to 236 mph (380 km/h).

△ **High-speed rail in France**
The TGV-PSE is a high-speed train built for operation between Paris and southeastern France. The original fleet had an orange and silver livery.

▷ **1991** Russia completes the Baikal-Amur Magistral—a major main line paralleling the classic Trans-Siberian.

▷ **1991** Germany enters the public high-speed rail era with the InterCityExpress (ICE).

▷ **1992** "Tram-train" services are launched in Karlsruhe, Germany. The new concept unites local rail and streetcars with vehicles that can run on both systems.

▷ **1993** Great Britain votes to privatize its railroads. In the years that follow, the state system is split up.

▷ **1994** In Germany, the former West German Deutsche Bundesbahn and East German Deutsche Reichsbahn are merged to form a new entity—Deutsche Bahn.

▷ **1994** The Channel Tunnel opens, connecting England and France by rail underneath the Dover Strait.

▷ **1995** China's Ji-Tong Railroad opens in Inner Mongolia. Known as the world's last steam main line, it is not fully converted to diesel until 2005.

◁ **A Union Pacific freight train** winding across the American landscape

High Speed Goes Global

Operating at speeds that were impossible on historic rail tracks, high-speed lines burst upon the world scene in 1964 with the introduction of the Shinkansen in Japan. In Europe, the French led the way, building a network of dedicated high-speed lines known as TGV (Train à Grande Vitesse), with the first route between Paris and Lyon opening in 1981. Spain's first high-speed line, the Alta Velocidad Española (AVE), opened between Madrid and Seville in 1992. The UK, with its Victorian rail network, lagged behind; despite the opening of the Channel Tunnel in 1994, it would not be until 2007 before the country's first dedicated high-speed railroad, HS1 was complete, ushering in high-speed rail travel between London and Paris.

▷ Soviet ER200, 1984

Wheel arrangement each car
2 x 4-wheel trucks

Power supply 3 kV DC overhead lines

Power rating 6-car set: 5,150 hp (3,840 kW)/14-car set: 15,448 hp (11,520 kW)

Top speed 124 mph (200 km/h)

Built of aluminum alloy in Riga, the ER200 is a Soviet high-speed train that was first introduced in 1984. At the time it was the first direct current (DC) intercity electric multiple-unit train with rheostatic braking. Later versions operate on the Moscow to St. Petersburg main line. Unit ER200-15 is on display at the Moscow Railway Museum.

△ AVE S-100, 1992

Wheel arrangement each car
2 x 4-wheel trucks

Power supply 3 kV DC overhead supply/25 kV 50 Hz AC overhead supply

Power rating 11.796 hp (8,800 kW)

Top speed 186 mph (300 km/h)

The Alta Velocidad Española (AVE) is a network of high-speed railroads operated in Spain by Renfe Operadora. It was Europe's longest high-speed network and, after China, the world's second longest. The first line between Madrid and Seville opened in 1992 using S-100 dual-voltage, electric multiple units built by Alstom.

▷ Thalys PBKA, 1996

Wheel arrangement 2 power cars + 8 passenger cars

Power supply 3 kV DC overhead supply/25 kV 50 Hz AC overhead supply/15 kV 16²/₃ Hz AC overhead supply/1,500 V DC overhead supply

Power rating 4,933 hp (3,680 kW) - 11,7961 hp (8,800 kW)

Top speed 186 mph (300 km/h)

Built by GEC-Alstom in France, the Thalys PBKA is a high-speed international train service, introduced in 1996, that can operate on four different electrical systems in France, Germany, Switzerland, Belgium, and the Netherlands. The 17 train sets built operate services between Paris, Brussels, Cologne (Köln), and Amsterdam, hence PBKA.

TECHNOLOGY

Transrapid Prototype

Developed in Germany, this high-speed monorail train with no wheels, gear transmissions, or axles, and has no rails or overhead power supply. Instead it levitates, or hovers, above a track guideway using attractive magnetic force between two linear arrays of electromagnetic coils, hence its name, "Maglev." Based on a patent from 1934, planning for it began in 1969 and the test facility was completed in 1987. The latest version Maglev 09 can cruise at over 300 mph (482 km/h). The only commercial application to date opened in China in 2002 and operates between Shanghai and its Pudong International Airport.

Revolutionary technology The two-car Maglev Transrapid prototype is seen in action at the test facility at the Emsland test track in Germany in 1980.

▷ Eurostar Class 373/1, 1993

Wheel arrangement each car
2 x 4-wheel trucks

Power supply 25 kV 50 Hz AC overhead supply/3,000 V DC overhead supply/1,500 V DC overhead supply/750 V DC third-rail (not used)

Power rating 4,600 hp (3,432 kW) - 16,360 hp (12,200 kW)

Top speed 186 mph (300 km/h)

Introduced in 1993, the Class 373/1 multi-voltage electric multiple units are operated by Eurostar on the high-speed line between London, Paris, and Brussels via the Channel Tunnel. In the UK these trains operated on the third-rail network to London's Waterloo Station until the completion of the HS1 line in 2007.

▷ **SNCF LGV Sud-Est TGV, 1981**

Wheel arrangement each car 2 x 4-wheel trucks

Power supply 1,500 V DC overhead supply/
25 kV 50 Hz AC overhead lines

Power rating 4,157 hp (3,100 kW) – 9,119 hp (6,800 kW)

Top speed 186 mph (300 km/h)

The French Train à Grande Vitesse (TGV) was
originally designed to be powered by gas turbines,
but the oil crisis of 1973 led to the first prototypes
being electrically powered. Built by GEC-Alstom,
the first of these dual-voltage high-speed trains
entered service on the LGV (Ligne à Grande Vitesse)
Sud-Est line between Paris and Lyon in 1981.

▷ **SJ X2, 1989**

Wheel arrangement each car 2 x 4-wheel bogies

Power supply 15 kV 16²⁄₃ Hz AC overhead lines

Power rating 4,370 hp (3,260 kW)

Top speed 124 mph (200 km/h)

Built of corrugated stainless steel, the Swedish railroad's
(Statens Järnvägar, or SJ) X2 high-speed tilting train
is designed to operate on the country's existing rail
network. In tests, it has reached 171 mph (276 km/h).
One train set was exported to China and others loaned
to Amtrak in the US and Countrylink in Australia.

▷ **DB ICE 1, 1991**

Wheel arrangement each car
2 x 4-wheel trucks

Power supply 15 kV 16³⁄₄ Hz AC,
overhead supply

Power rating 5,094 hp (3,800 kW)–
6,437 hp (4,800 kW)

Top speed 174 mph (280 km/h)

Introduced in 1991, InterCityExpress
(ICE) 1 was Germany's first truly high-
speed public train. Sixty train sets
were built, each one consisting of
a power car at each end and either
12 or 14 passenger cars; a 12-car set
can accommodate 743 passengers.

Building Great Railroads
Eurostar

The modern era of high-speed rail travel between London, Brussels, and Paris began with the opening of the Channel Tunnel in 1994. However, on the English side of the channel, the new Eurostar trains were forced to run on a Victorian rail system until the completion of the HS1 link in 2007.

London ⬤▥▥▥▥▥▥

London St. Pancras 1
After initially operating from London Waterloo from 1994, the Eurostar terminal was relocated to the refurbished London St. Pancras International in 2007 with the opening of the HS1 link.

THE IDEA OF A TUNNEL under the English Channel to link the UK and France was not new. Various proposals were made during the 19th and early 20th centuries, but British fears that a tunnel could be used by an army invading England scuppered most plans, even though one 1929 design included a system for flooding the tunnel to repel invaders.

It was not until the 1960s that the French and British governments agreed to a modern project. Construction finally began in 1974, but halted within a year when the British, seeing costs soar and the economy crumble, canceled the project. Eventually, in 1986, a private consortium of British and French banks and construction firms agreed to build the tunnel, work beginning from both sides in 1988. Two years later, the two ends of the service tunnel met under the channel.

Opened in 1994, the Channel Tunnel extends for 31⅓ miles (50.45 km) between Folkestone in England and Coquelles, near Calais in France. It consists of two single-track rail tunnels separated by a service tunnel, which can be used for passenger evacuation in an emergency. It is not a rail tunnel but a roadway where the tunnel maintenance crews use zero-emissions electric vehicles.

The new Eurostar service between Paris and London demanded high-speed rail lines, and the French were first out of the gate, opening the

Eurostar at St. Pancras International
St. Pancras took over from Waterloo as London's international terminal in 2007, with the inauguration of the high-speed HS1 link from the Channel Tunnel.

1929 POSTER FOR PRE-EUROSTAR SERVICES

207-mile (333-km) LGV Nord in 1993. This electrified line connects Gare du Nord, Paris to the Belgian border and the Channel Tunnel via Lille. The Belgians followed with their 55-mile (88-km) HSL 1, which opened in 1997, linking LGV Nord to Brussels-South.

In England, the Eurostar trains ran at lower speeds on existing tracks between Folkestone in Kent and into special platforms at London Waterloo, a busy commuter station. Services began to Gare du Nord in Paris and Brussels-South station on November 14, 1994. It was to be a further 13 years before Britain's new 67-mile (108-km) High Speed 1 (HS1) line between Folkestone and the newly refurbished St. Pancras International station in London opened on November 14, 2007. This reduced the travel time between London and Brussels to 1 hour 51 minutes. London to Paris took just 2 hours 15 minutes, more than four hours faster than when passengers had to disembark, cross the channel by ferry, and then board another train bound for the capital on the French side.

Speed restrictions
The series 373000 TGV (BR Class 373 in the UK) reaches high speeds in open country, but is restricted to 99 mph (160 km/h) in the Channel Tunnel.

KEY FACTS

DATES
1988 February: Channel Tunnel building and tunneling begins
1990 December: the French and British tunnels meet underground
1993 June: the first Eurostar test train travels through the tunnel from France to the UK

TRAIN
Train set *Inter-Capital* (31 sets built, 27 in Eurostar service): 18 passenger cas; 1,293 ft (394 m) long, capacity 750
Train set *North of London* or *Regional* (7 sets built, on long-term lease to SNCF): 14 passenger cars; 1,050 ft (320 m) long, capacity 558
Train set *Nightstar* International service intended to run beyond London. Canceled 1999—all 139 coaches sold to Via Rail in Canada
Locomotives 27 Eurostar electric multiple unit (EMU) sets currently in service, Class 373/1 (UK) and TGV373000 (France). 2 power cars per set
Carriages 3 Eurostar travel classes: business premiere, standard premiere, and standard
Speed 186 mph (300 km/h) on high-speed lines; 99 mph (160 km/h) in the Channel Tunnel

JOURNEY
London St. Pancras to Gare du Nord, Paris
305 miles (492 km); 2 hours 15 minutes (from 2007)
London St. Pancras to Brussels-South
232 miles (373 km); 1 hour 51 minutes (from 2007)

RAILROAD
Gauge Standard gauge 4 ft 8 ½ in (1.435 m), cleared to larger European loading gauge
Channel Tunnel World's second-longest tunnel at 31 ⅓ miles (50.45 km) and longest undersea rail tunnel in the world at 23 ½ miles (37.9 km)
Bridges Medway Viaduct, UK, 4,265 ft (1.3 km)
Lowest point 250 ft (76 m) below sea level

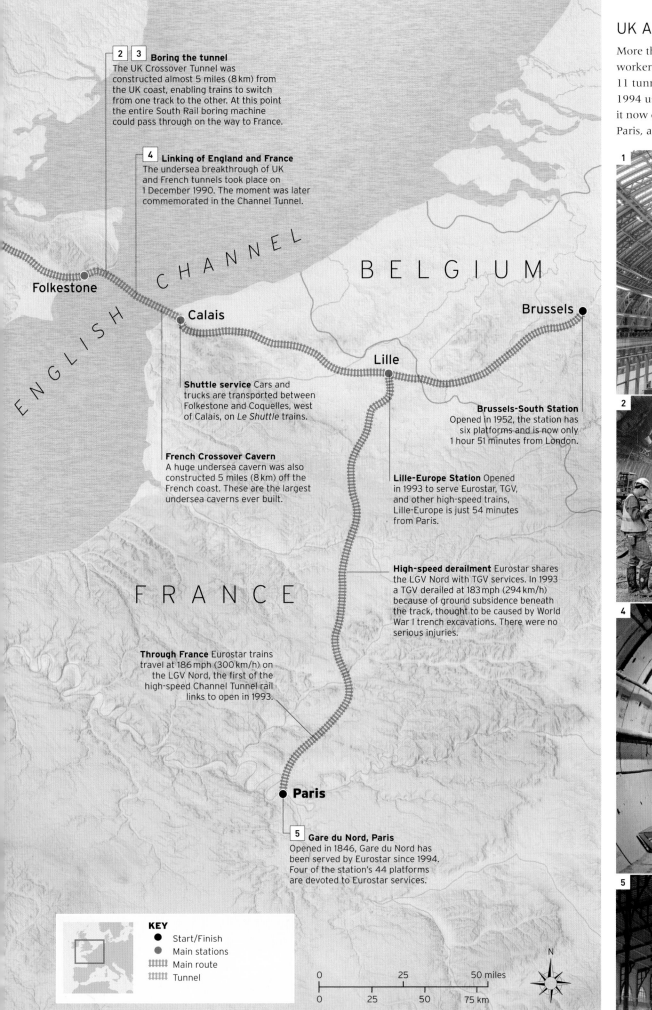

2 3 Boring the tunnel
The UK Crossover Tunnel was constructed almost 5 miles (8 km) from the UK coast, enabling trains to switch from one track to the other. At this point the entire South Rail boring machine could pass through on the way to France.

4 Linking of England and France
The undersea breakthrough of UK and French tunnels took place on 1 December 1990. The moment was later commemorated in the Channel Tunnel.

ENGLISH CHANNEL

BELGIUM

Folkestone

Calais

Brussels

Lille

Shuttle service Cars and trucks are transported between Folkestone and Coquelles, west of Calais, on *Le Shuttle* trains.

Brussels-South Station
Opened in 1952, the station has six platforms and is now only 1 hour 51 minutes from London.

French Crossover Cavern
A huge undersea cavern was also constructed 5 miles (8 km) off the French coast. These are the largest undersea caverns ever built.

Lille-Europe Station Opened in 1993 to serve Eurostar, TGV, and other high-speed trains, Lille-Europe is just 54 minutes from Paris.

High-speed derailment Eurostar shares the LGV Nord with TGV services. In 1993 a TGV derailed at 183 mph (294 km/h) because of ground subsidence beneath the track, thought to be caused by World War I trench excavations. There were no serious injuries.

FRANCE

Through France Eurostar trains travel at 186 mph (300 km/h) on the LGV Nord, the first of the high-speed Channel Tunnel rail links to open in 1993.

Paris

5 Gare du Nord, Paris
Opened in 1846, Gare du Nord has been served by Eurostar since 1994. Four of the station's 44 platforms are devoted to Eurostar services.

KEY
● Start/Finish
● Main stations
▒ Main route
▒ Tunnel

0   25   50 miles
0   25   50   75 km

N

UK AND FRANCE UNITED

More than 13,000 engineers, technicians, and workers labored to link the UK with France; and 11 tunnel-boring machines were used. Since 1994 use of the tunnel has grown steadily, and it now carries more passengers between London, Paris, and Brussels than all airlines combined.

Diesel's Next Generation

By the early 1980s, the first home-grown generation of diesel–electric locomotives in Europe and North America had reached the end of their working lives. US engine builders General Electric and the General Motors EMD brand then began to dominate the scene on both continents with their highly successful, more powerful and efficient heavy-freight machines, which remain in operation today. In the UK, on the other hand, diesel locomotive building ended completely in 1987 when the last engine, BR Class 58 diesel-electric No. 58 050, rolled off the production line at the famous Doncaster Works.

△ **BR Class 58, 1984**

Wheel arrangement	Co-Co
Transmission	electric
Engine	Ruston Paxman 12-cylinder diesel
Total power output	3,300 hp (2,460 kW)
Top speed	80 mph (129 km/h)

Designed with an optimistic eye on export potential, 50 of the Class 58 heavy-freight, diesel-electric locomotives were built by British Rail Engineering Ltd. in Doncaster between 1983 and 1987. They had a short working life in Great Britain, with the last retired in 2002. Since then, 30 have been leased by railroads in the Netherlands, France, and Spain.

△ **Amtrak GE Genesis, 1992**

Wheel arrangement	B-B
Transmission	electric
Engine	General Electric V12 or V16 4-stroke supercharged diesel
Total power output	4,250 hp (3,170 kW)
Top speed	110 mph (177 km/h)

General Electric Transportation Systems built 321 of these low-profile, lightweight, diesel-electric locomotives between 1992 and 2001. They operate most of Amtrak's long-haul and high-speed rail services in the US and Canada. A dual-mode version can also use 750 v DC current from third-rail in built-up areas such as New York.

△ **IÉ Class 201, 1994**

Wheel arrangement	Co-Co
Transmission	electric
Engine	EMD V12 2-stroke diesel
Total power output	3,200 hp (2,386 kW)
Top speed	102 mph (164 km/h)

Thirty-two of these powerful diesel–electric locomotives were built by General Motors in Ontario, Canada, for Iarnród Éireann in Ireland between 1994 and 1995. Two were also built for Northern Ireland Railways. They are all named after Irish rivers and operate on the Dublin to Cork express trains and on the *Enterprise* between Dublin and Belfast.

▷ **UP GM EMD Class SD60, 1984**

Wheel arrangement	C-C
Transmission	electric
Engine	EMD 16-cylinder diesel
Total power output	3,800 hp (2,834 kW)
Top speed	65 mph (105 km/h)

Built by General Motors, the heavy freight EMD Class SD60 diesel-electric locomotive was introduced in 1984. Production ceased in 1995, by which time 1,140 had been delivered to nine US railroads, Canadian National Railways, and Brazil. Union Pacific Railroad bought 85 of the SD60, seen here, and 281 of the SD60M variant.

◁ **BR GM EMD Class 66, 1998**

Wheel arrangement Co-Co

Transmission electric

Engine EMD V12 two-stroke diesel

Total power output 3,000 hp (2,238 kW)

Top speed 75 mph (121 km/h)

A total of 446 of these diesel-electric freight locomotives were built by Electro-Motive Diesel in the US for British railroads between 1998 and 2008. Over 650 of this highly successful design have also been sold to several European freight operators as well as the Egyptian state railroads.

▷ **DWA Class 670 railcar, 1996**

Wheel arrangement 2-axle

Transmission mechanical

Engine MTU 6V 183 TD 13 diesel

Total power output 335 hp (250 kW)

Top speed 62 mph (100 km/h)

Incorporating parts used in buses, six of these double-deck diesel railcars were built by German Wagon AG (DWA) for German state railroads in 1996 after a prototype was unveiled in 1994. A number remain in service.

◁ **ADtranz DE AC33C, 1996**

Wheel arrangement Co-Co

Transmission electric

Engine General Electric V12 diesel

Total power output 3,300 hp (2,462 kW)

Top speed 75 mph (121 km/h)

Equipped with General Electric diesel engines, these powerful locomotives, nicknamed "Blue Tigers," were built by German manufacturer ADtranz between 1996 and 2004. Eleven units, including No. 250 001-5, seen here, were made for leasing in Germany, while Pakistan Railways ordered 30 and Keretapi Tanah Melayu in Malaysia bought 20.

▷ **HSB Halberstadt railcar, 1998**

Wheel arrangement 2 x 4-wheel trucks (1 powered)

Transmission mechanical

Engine Cummins 6-cylinder 10.8 litre diesel

Total power output approx 375 hp (280 kW)

Top speed 31 mph (50 km/h)

Four of these were built in 1999 by the Halberstadt Works, then part of Deutsche Bahn, for the Harzer Schmalspurbahnen (Harz Narrow-Gauge Railway). They still work services at times, running on lines that are lightly used.

A New Wave of Electrics

The demise of steam power in Western Europe during the 1950s and 1960s saw the spread of electrification across much of the continent. The soaring price of oil in the 1970s added further impetus for national railroads to switch from hurriedly introduced diesel locomotives to electric haulage. However, the power supplies varied greatly from country to country, and with the growth of transnational railroad freight services, a new generation of multivoltage electric locomotives had started to appear by the 1990s.

▷ CSD Class 363, 1980

Wheel arrangement	B-B
Power supply	25 kV 50 Hz AC/3,000 V DC, overhead supplies
Power rating	4,102–4,666 hp (3,060–3,480 kW)
Top speed	75 mph (120 km/h)

The prototype Class 363 dual-voltage locomotive was built by Skoda Works for the Czechoslovakian state railroads. It was the first multi-system electric engine in the world equipped with power thyristor pulse regulation and has a distinctive sound in three frequencies when accelerating.

◁ DR Class 243, 1982

Wheel arrangement	Bo-Bo
Power supply	15 kV 16.7 Hz AC, overhead supply
Power rating	4,990 hp (3,721 kW)
Top speed	75 mph (120 km/h)

Over 600 of these mixed-traffic electric locomotives were built by L.E.W. Hennigsdorf for the Deutsche Reichsbahn between 1982 and 1991. Originally classified as DR Class 243, they became Class 143 under the renumbering program that followed Germany's reunification.

Glacier Express

Named in honor of the Rhone Glacier, which it passed at the Furka Pass, the Glacier Express was introduced between St. Moritz and Zermatt, Switzerland, on June 25, 1930. It was originally operated by three 3-ft 3-in- (1-m-) gauge railroad companies, the Brig-Visp-Zermatt Bahn (BVZ), the Furka Oberalp Bahn (FO), and the Rhaetian Railway (RhB). While two of the lines were electrified, steam locomotives were used on the FO section until 1942, when that line was also electrified. It runs daily all year round but is not exactly an "express," since it takes 7½ hours to cover 181 miles (291 km), much of it on a rack-and-pinion system. Since 2008 much of the route on the Albula and Bernina railroads has been declared a UNESCO World Heritage Site.

Scenic ride The train passes through stunning Alpine scenery, crossing 291 bridges, burrowing through 91 tunnels, and gaining height on numerous spirals.

△ PKP Class EP09, 1986

Wheel arrangement	Bo-Bo
Power supply	3,000 V DC, overhead supply
Power rating	3,914 hp (2,920 kW)
Top speed	99 mph (160 km/h)

A total of 47 of the Class EP09 express passenger electric engines were built by Pafawag of Wroclaw for the Polish state railroads between 1986 and 1997. First entering service in 1988, they operate trains on main lines from Warsaw and Kraków.

△ SNCF Class BB 26000, 1988

Wheel arrangement	B-B
Power supply	25 kV AC/1,500 V DC, overhead supplies
Power rating	7,500 hp (5,595 kW)
Top speed	124 mph (200 km/h)

These multipurpose, dual-voltage electric engines were constructed for the French state railroads between 1988 and 1998; a total of 234 were built. A further 60 triple-voltage locomotives, which were made between 1996 and 2001, are classified as SNCF Class BB 36000.

△ BR Class 91, 1988

Wheel arrangement Bo-Bo

Power supply 25 kV AC, overhead supply

Power rating 6,480 hp (4,832 kW)

Top speed 125 mph (204 km/h)

Delivered between 1988 and 1991, 31 of the Class 91 express locomotives were built at Crewe Works for British Rail. Designed to reach 140 mph (225 km/h) but now only used at 125 mph (204 km/h), they operate express trains in a push-pull mode on the East Coast Main Line between London King's Cross and Edinburgh.

△ FS Class ETR 500, 1992

Wheel arrangement power cars: 2 x 4-wheel motorized trucks

Power supply 3 kV DC, overhead supply

Power rating complete train: 11,796 hp (8,800 kW)

Top speed 155 mph (250 km/h)

Following four years of testing, 30 Class ETR 500 high-speed, single-voltage electric trains were introduced on the Italian state railroads between 1992 and 1996. Before the production models were constructed, a prototype motor car was built and tested. Coupled to an E444 locomotive on the Diretissima Line between Florence and Rome, it attained a speed of 198 mph (319 km/h) in 1988.

◁ SBB Cargo Bombardier Traxx, 1996

Wheel arrangement Bo-Bo

Power supply 15 kV 16.7 Hz AC/25 kV 50 Hz AC, overhead supply

Power rating 7,500 hp (5,595 kW)

Top speed 87 mph (140 km/h)

From 1996 the Bombardier Traxx, dual-voltage electric locomotives were introduced on many European railroads. Since then, around 1,000 have been built at the company's assembly plant in Kassel, Germany, of which 35 of the F140 AC variant, seen here, are operated by SBB Cargo in Switzerland.

▽ BR Class 92, 1993

Wheel arrangement Co-Co

Power supply 25 kV AC, overhead supply/750 V DC third-rail

Power rating 5,360–6,760 hp (3,998–5,041 kW)

Top speed 87 mph (140 km/h)

Designed to haul freight trains through the Channel Tunnel between England and France, the 46 Class 92, dual-voltage electric locomotives were built by Brush Traction and ABB Traction and assembled at Brush Traction's erecting shops in Loughborough, England, between 1993 and 1996. They are operated by GB Railfreight/Europorte 2 and DB Schenker.

▷ Amtrak Class HHP-8, 1999

Wheel arrangement B-B

Power supply 12.5 kV 25 Hz AC/12.5 kV 60 Hz AC/ 25 kV 60 Hz AC, overhead supplies

Power rating 8,000 hp (5,968 kW)

Top speed 125 mph (201 km/h)

Fifteen of these express passenger electric locomotives were built for Amtrak by Bombardier and Alstom in 1999. The Amtrak locomotives hauled trains on the Northeast Corridor between Washington, DC and Boston until they were retired in 2012.

Palace on Wheels

Traveling in the style of a maharaja through India's most evocative destinations is one of life's most luxurious railroad experiences. The *Palace on Wheels*, one of the world's top five luxury trains, is a hotel on rails outfitted to resemble the Indian stately cars from a bygone age. Passengers travel in private saloons with personal attendants on a tour of India's Golden Triangle of historic sites that ends with a trip to the Taj Mahal.

THE CARS of the *Palace on Wheels* offer passengers the experience of stepping back in time. The train has revived the look and feel of the magnificent railroad coaches that once transported the rulers of Rajasthan and Gujarat, the Nizams of Hyderabad, and the Viceroys of India during the time of the British Empire.

After India gained its independence from the UK in 1947, the "royal" rail cars were taken out of commission, as their extravagantly decorated interiors made them inappropriate for India's fleet of standard passenger trains. However, in 1982, Indian Railways teamed up with the Rajasthan Tourist Development Cooperation and brought these plush vehicles back into service by introducing the *Palace on Wheels* luxury train experience. The seven-day round trip, which starts in New Delhi and tours northwest India, has since become a tourist attraction that draws people from around the world.

The coaches of the *Palace on Wheels* have been revamped more than once in the last 30 years. Currently, the train is made up of 14 saloons, a kitchen car, two restaurants, a bar with a lounge, and four service cars. Each saloon is named after one of the erstwhile princely states, and their interiors reflect the history of the region through paintings, furniture, handicrafts, and furnishings. To add a further touch of majesty to the experience, the train also offers personal "Khidmatgars," or attendants, who are available to serve guests around the clock.

STEAM ENGINE

DIESEL ENGINE

Aspiring to royalty
The *Palace on Wheels* train is a spectacular recreation of the royal and official trains of the Indian 3-ft 3-in (1-m-) gauge in the 1920s and 1930s. Today, its cars are rarely hauled by steam—most journeys are powered by a diesel locomotive.

SPECIFICATIONS FOR CARS	
Origin	India
In-service	1982-present
Coaches	14
Passenger capacity	approx. 80
Route	Rajasthan and the Golden Triangle (Delhi-Jaipur-Agra)

Trip of a lifetime
The week-long trip on the *Palace on Wheels* takes passengers through northwestern India on a nostalgic journey to some of the most popular tourist spots in the Golden Triangle.

Windows run the full length of the car

Exterior paintwork is identical on every saloon car

Coat of arms identifies the princely state that inspired the interior decoration

Painted sign shows name of saloon car

Passenger doors at each end with stylized oval windows

KISHANGARH SALOON CAR

Fit for a king
The two restaurants on board the *Palace on Wheels* are called *Maharaja* (shown here) and *Maharani*. An original royal dining saloon is kept in the National Railway Museum in Delhi.

JAIPUR SALOON AND BEDROOM

The Jaipur saloon is decorated in colors that represent the former Rajput state of Jaipur, while the exterior of the car bears its coat of arms. The ceiling is adorned in the region's famed "phad" (foil work), and illustrates religious festivals such as Teej, Holi, Gangaur, and Diwali. Each saloon consists of four coupes (sleeping rooms) and a bathroom. A mini pantry and a lounge provide additional comfort.

1. Name of car embossed on metal plate **2.** "Phad" (foil work) on ceiling depicting festivals celebrated in Rajasthan **3.** Glass and gilt ceiling light **4.** Saloon with banquet-style sofas and painted fresco ceiling **5.** Metal hand plate on door **6.** Car corridor **7.** Coupe (sleeping room) **8.** Mirror inside the coupe **9.** Switches for lights and music **10.** Private bathroom with elegant modern fixtures and mirror

PALACE ON WHEELS BAR

The lounge bar is designed to reflect a contemporary royal style with flourishes that hark back to the Rajput era. With its wood, marble, and brass fixtures, the bar area epitomizes the aesthetic of the time. A selection of antique pitcher designs ornaments the front of the counter area, depicting some of the drink-pouring vessels the maharajas would have used.

11. Bar and lounge car **12.** Marble-top bar counter **13.** Antique pitcher design in marble, with gold inlay work, on front of bar counter **14.** Emergency stop chain **15.** Chandelier **16.** Peacock motif in tinted glass **17.** "Jaahli" (teak latticework) panel **18.** Armchairs with "patra" (oxidized white metal) work on borders **19.** Deep-cushioned sofa with raw silk upholstery **20.** Intricately carved elephant head design—a symbol of prosperity—at end of armrest

MAHARANI RESTAURANT

The Rajasthani theme continues in the interior design of the Maharani (meaning "Queen") dining cabin, with floral carpets and curtains, and featuring framed art from the Mughal period hanging on the walls. The most opulent touch is arguably the mirrored and teakwood-paneled ceiling.

1. Sumptuous dining room with mirrored ceiling **2.** Mughal art in marble, created with vegetable colors, on wall **3.** Silk-embroidered drapes with floral design **4.** Tree motif in stained glass on restaurant door **5.** Paneled corridor **6.** Kitchen positioned at one end of restaurant car

MAHARAJA RESTAURANT

Maharaja means "King" in Hindi, and accordingly, this dining carriage has a more masculine feel compared to the Maharani. Drapes of royal blue adorn the elegant, mahogany-dominated decor. The seating is arranged in groups of four. Both restaurants serve different varieties of cuisine, although there is an emphasis on Rajasthani dishes.

7. Name plaque above the door **8.** Air vent in central ceiling panel **9.** Wall light with painted glass shade **10.** Gold-embroidered "zari" work on velvet drape **11.** Restaurant car decorated with mahogany paneling

ROYAL SPA

The *Palace on Wheels* boasts a car dedicated to spa services. It is the most recent addition to the luxury experience. Although still majestic, the decor of the treatment rooms is toned down to encourage maximum customer relaxation. The fully equipped spa offers massages, treatments, and various revitalizing solutions.

12. Corridor in the Royal Spa **13.** Double-bed massage suite **14.** Reclining chair and sink **15.** Pedicure bowl with rose petals

GENERATOR CAR AND CONDUCTOR'S COMPARTMENT

The conductor's compartment and generator car are located at the front of the train, away from the palatial setting of the passenger cars. The generator provides electricity to power the lights, appliances, kitchen, and bar equipment. In the conductor's cabin, a close eye is kept on gauge and meter readings to ensure the train runs smoothly and that passengers have a comfortable trip.

16. Power control panel in the generator car **17.** Conductor's compartment **18.** Handbrake **19.** Temperature control panel **20.** Vent control **21.** Airbrake

Urban Rail Solutions

While pioneering urban railroads such as London's Metropolitan Railway and Chicago's South Side Elevated Railroad originally ran on steam, by the late 1930s, electrified railroads such as the Budapest Metro, Moscow Metro, and London Underground were carrying huge numbers of commuters between their suburban homes and downtown offices. With the world's cities still expanding during the late 20th century, modern electrically powered rapid transit systems (RTS) such as streetcars and surface and underground railroads, many using driverless automatic trains, were built to transport millions of passengers each day, very quickly and over short distances.

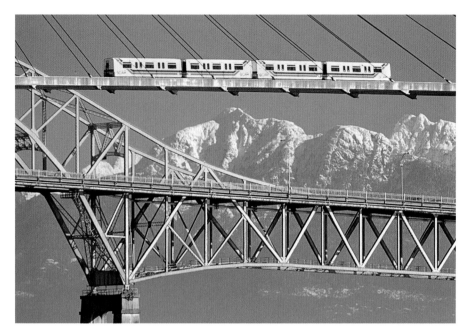

▷ Vancouver SkyTrain RTS ICTS Mark I, 1985

Wheel arrangement	2-car sets
Power supply	750 V DC, third rail
Power rating	120 hp (88 kW) x 4 per 2-car set
Top speed	50 mph (80 km/h)

The 43-mile (69-km) Vancouver SkyTrain is an RTS serving Vancouver and its suburbs. Trains on the Expo Line and Millennium Line are automated and are driven by linear induction motors. The cars run in two- to six-car configurations. This is a four-car Mark I train of the Intermediate Capacity Transit System (ICTS) built by the Urban Transportation Development Corporation of Ontario.

△ SDTI Duewag U2 cars, 1980/81

Wheel arrangement	double-ended, 6-axle, articulated
Power supply	600 V DC, overhead supply
Power rating	480 hp (300 kW)
Top speed	50 mph (80 km/h)

The San Diego Trolley is a 53-station, three-route light rail system in the city of San Diego, CA, which opened in 1981. The articulated cars initially used were U2 vehicles built in Germany by Duewag. These cars also worked in Edmonton and Calgary, Canada, and in Frankfurt, Germany.

△ Berlin *U-Bahn* F-type train, 1992/1993

Wheel arrangement	2-car sets
Power supply	750 V DC, third rail
Power rating	734 hp (540 kW) per 2-car set
Top speed	45 mph (72 km/h)

The 152-mile (245-km) Berlin *U-Bahn* (or underground railroad) opened in 1902 and, despite problems caused by the division of Berlin during the Cold War, today serves 170 stations across 10 lines. The system uses both *Kleinprofil* (small profile) trains and *Grossprofil* (large profile) trains, such as the F-type. The trains are worked in four-, six-, and eight-car configurations.

▽ Vienna ULF tram, 1998

Wheel arrangement 2- or 3-car articulated

Power supply 216–480 kW (289–643 hp)

Power rating 653 hp (480 kW)

Top speed 50 mph (80 km/h)

The Ultra Low Floor (ULF) cars, built by the consortium of Siemens of Germany and Elin of Austria, were introduced on Vienna's tram network in 1998 and in 2008 in Oradea, Romania. With a floor only 7 in (18 cm) above the ground, they provide easy access for wheelchairs or strollers.

△ T&W Metro, 1980

Wheel arrangement 2-car articulated (6-axle articulated sets)

Power supply 1,500 V DC, overhead supply

Power rating 410 hp (301.5 kW)

Top speed 50 mph (80 km/h)

The 46-mile (74-km) Tyne & Wear Metro in Newcastle-upon-Tyne in northeast England is a hybrid light railroad system with suburban, interurban, and underground sections. A total of 90 two-car articulated sets, usually coupled together in pairs, were built between 1978 and 1981 by Metro Cammell in Birmingham.

▽ Luas Alstom Citadis tram, 1997

Wheel arrangement 3-, 5-, and 7-car articulated (8-, 12-, and 16-axle articulated sets)

Power supply 750 V DC, overhead lines

Power rating 979 hp (720 kW)

Top speed 44 mph (70 km/h)

The Citadis is a family of low-floor trams built by Alstom in France and Spain and popular in many cities around the world. The 23-mile (37-km) Luas tram system in Dublin, Ireland, uses the three-car 301 and five-car 401 variants on the city's Red Line, while the seven-car 402 variant works on the Green Line.

△ SMRT North–South Line C151, 1987

Wheel arrangement 6-car sets

Power supply 750 V DC, third rail/ 1,500 V DC, overhead supply

Power rating 2,937 hp (2,160 kW)

Top speed 50 mph (80 km/h)

Singapore's Mass Rapid Transit system started life when the North–South Line opened in 1987. Since then, it has been extended to 93 miles (150 km), serving 106 stations on five routes. Six-car C151 (shown), C151A, and C751B trains draw current from a third rail and have an automatic train operation system. C751A trains, which are fully automatic and driverless, use an overhead supply.

▽ Gatwick Adtranz C-100, 1987

Wheel arrangement 2-car sets with rubber tires

Power supply 600 V AC

Power rating 100.5 hp (75 kW) per car

Top speed 28 mph (45 km/h)

This elevated, fully automatic, driverless, guided people-mover system began operation in 1987. The train connects the North and South Terminals at London's Gatwick Airport, a distance of $^3/_4$ mile (1.2 km). Similar systems have proved popular in airports and cities around the world.

AFTER 2000
RAIL REVIVAL

RAIL REVIVAL

As the world looks for alternatives to roads, rail has once again become a priority. New energy-efficient locomotives have been launched and the search for even higher speeds continues. In 2003 an experimental Maglev (*mag*netic *lev*itation) train in Japan reached speeds of 361mph (581km/h), and in 2004 the world's first commercial high-speed Maglev began operations in Shanghai, China. Two years later, the opening of the railroad to Lhasa, Tibet, broke a different kind of record, taking normal trains to altitudes never before reached. Oxygen masks are available to passengers on trains that operate at more than 16,000 ft (5,000m). In 2007 a specially adapted TGV in France broke the record for a conventional train.

Expansion of high-speed rail has continued around the world; China has opened thousands of miles of track to become the home of the world's largest high-speed network, and Spain, which entered the new era in 1992, has set out to create Europe's biggest system. The launch of the *Acela Express* in the US pushed the nation's maximum speed up to 150mph (241km/h), while the UK completed a dedicated line connecting London and the Channel Tunnel (HS1).

Yet progress is not all about going faster. Metros and light rail have continued to expand their reach—and operators have pressed ahead with greater automation. As the 21st century's first decade drew to a close, the Dubai Metro became the world's longest fully automated line, at 47 miles (75 km).

Rail travel has reinvented itself as a luxurious alternative to cramped aircraft or gridlocked roads - passengers can travel through ever-changing landscapes in style. More than two centuries after it began, the railway era is far from over.

> " Any railway, **working properly,** is a marvel of **civilized** cooperation "
>
> LIBBY PURVES, *THE TIMES* [UK], MAY 14, 2002

Key events

▷ **2000** The *Acela Express* is introduced in the US. It reaches speeds of up to 150 mph (241km/h).

▷ **2003** The first section of the UK's high-speed Channel Tunnel Rail Link is opened. It reaches London in 2007 and is now known as High Speed 1 (HS1).

▷ **2003** An experimental Maglev train reaches 361mph (581 km/h) in Japan, a new world record for a manned train.

▷ **2004** The world's first commercial high-speed maglev system opens in Shanghai, China.

△ **Shanghai Transrapid maglev train**
The world's only maglev train service runs between Shanghai and the city's Pudong International Airport. It reaches speeds of 268mph (431km/h).

▷ **2006** Services start on the world's highest conventional railroad. The route in Tibet reaches up to 16,640 ft (5,072 m) above sea level.

▷ **2007** An experimental French TGV set sets a new world record for conventional-wheeled trains of nearly 357 mph (575 km/h).

▷ **2007** China enters the modern high-speed rail age with a new dedicated line.

▷ **2009** The first section of Dubai's metro opens, followed by another in 2010; at 47 miles (75 km), it is the world's longest fully automated metro.

▷ **2012** Completion of the major sections of the Beijing to Hong Kong high-speed line makes it the world's longest. When it is finished in 2015, the line will be more than 1,450-miles (2,334-km) long.

THE VENICE SIMPLON–ORIENT-EXPRESS
...A JOURNEY LIKE NO OTHER

vsoe.com

△ **Orient Express luxury travel**
This modern poster for the Venice Simplon–Orient Express features a liveried porter and hints at a return to the luxury travel of a bygone era.

◁ **A "Bullet Train"** speeds through a district of high-rise office buildings in Tokyo, Japan

Universal Applications

The beginning of the 21st century saw changes in the way manufacturers dealt with their customers. Instead of railroad companies telling the equipment manufacturer exactly what they wanted built, the manufacturers started offering railroad operators product ranges based on universal "platforms," much like the auto or aviation industries. As a result, some commuters in California now travel in trains similar to those in Berlin or Athens, and interoperable locomotives, able to run on multiple traction voltages and using several different signaling systems, are now common in Europe.

△ **Siemens Eurosprinter ES64 U2/U4, 2000**

Wheel arrangement	Bo-Bo
Power supply	1,500 V DC/3,000 V DC and 15 kV AC/25 kV AC, overhead lines
Power rating	8,579 hp (6,400 kW)
Top speed	143 mph (230 km/h)

Siemens introduced the Eurosprinter family of locomotives following big orders from Deutsche Bahn in Germany and ÖBB in Austria. The Eurosprinter range has four basic body shells and multiple versions. The ES64 U4 (EuroSprinter 6400 kW Universal 4 system) is the most flexible and is able to operate in multiple countries.

△ **Siemens Desiro Classic, 2000**

Wheel arrangement	2-car DMU
Transmission	mechanical
Engine	2 x MTU 1800 6R
Total power output	845 hp (630 kW)
Top speed	74 mph (120 km/h)

Siemens has now sold over 600 of its first "Desiro" model, the two-car articulated diesel-powered Desiro Classic. The trains are used for regional passenger services in Europe, and the design has been exported outside Europe to southern California. Electric versions have also been built for Bulgaria, Greece, and Slovenia.

▽ **Voith Gravita, 2008**

Wheel arrangement	B-B
Transmission	hydraulic
Engine	8 V 4000 R43
Total power output	1,341 hp (1,000 kW)
Top speed	62 mph (100 km/h)

The Gravita family of locomotives, developed by Voith, is designed for freight traffic on lightly used lines. Germany's Deutsche Bahn purchased 130 locomotives of two types—99 of the Gravita 10BB and 31 of the more powerful 15BB model.

▷ Siemens Desiro-RUS, 2013

Wheel arrangement 2-car EMU

Power supply 3,000 V DC and 25 kV AC, overhead lines

Power rating 3,420 hp (2,550 kW)

Top speed 100 mph (160 km/h)

Desiro EMUs have been built for several countries from the UK to Slovenia, Greece, and Thailand. Russian railroads (RZD) ordered 38 Desiro-RUS to operate services at the 2014 Sochi Winter Olympics. The trains, branded *Lastochka* (meaning "swallow"), were built at Siemens' Krefeld factory in Germany.

◁ Bombardier ALP45 DP, 2012

Wheel arrangement Bo-Bo

Transmission electric

Engine 2 x Caterpillar 3512C

Power supply 25 kV and 12.5 kV AC, overhead wires

Total power output diesel: 4,200 hp (3,132 kW)/ electric: 5,362 hp (4,000 kW)

Top speed diesel: 100 mph (161 km/h)/ electric: 125 mph (203 km/h)

These engines were designed for through operation from busy electric lines to non-electrified regional routes in North America to facilitate "one-seat rides"—traveling without changing trains. The locomotive can switch from electric to diesel (and vice versa) while moving. Bombardier has sold 46 to New Jersey Transit and Agence Métropolitaine de Transport in Montreal, Canada.

▷ Voith Maxima, 2008

Wheel arrangement C-C

Transmission hydraulic

Engine ABC 16 V DZC

Total power output 4,826 hp (3,600 kW)

Top speed 74 mph (120 km/h)

Voith introduced its most powerful single-engine, diesel-hydraulic locomotive ever built for freight operators in Europe in 2008. Two versions are available—the Maxima 40CC and lower-powered Maxima 30CC. Around 20 have been sold, mostly to Germany-based operators.

△ Vossloh Eurolight, 2010

Wheel arrangement Bo-Bo

Transmission electric

Engine Caterpillar C175-16

Total power output 3,753 hp (2,800 kW)

Top speed 124 mph (200 km/h)

The Eurolight design from Vossloh aims to maximize available power while minimizing axle weight, which enables the locomotive to operate even on rural routes, often not built for heavy trains. By using a lighter-weight engine and lightweight body, the locomotive weighs under 90 tons (79 metric tons).

△ Vossloh G6, 2010

Wheel arrangement C

Transmission hydraulic

Engine Cummins QSK-23-L

Total power output 900 hp (671 kW)

Top speed 50 mph (80 km/h)

Vossloh builds the G6 diesel-hydraulic locomotive in Kiel in Germany at the former Maschinenbau Kiel (MaK) factory. So far it has been sold mostly to industrial operators in Germany. Verkehrsbetriebe Peine-Salzgitter (VPS) runs a large rail network serving the steel industry at Salzgitter (central Germany) and has bought 40 G6s to replace 43 older diesel switchers.

△ Alstom Prima II, 2010

Wheel arrangement Bo-Bo

Power supply 3,000 V DC and 25 kV AC, overhead lines

Power rating 5,630 hp (4,200 kW)

Top speed 124 mph (200 km/h)

Alstom developed the Prima II prototype in 2008 and has sold 20 to Moroccan railroads (ONCF). The locomotives are used for passenger trains on electrified routes. Able to work on all traction voltages, the Prima II model can be built as four-axle locomotives or six-axle freight versions.

Historic Railroads

Although many now operate solely for the benefit of the tourism industry, today's historic and heritage railroads were first created to fulfill specific industrial or commercial functions. While a few continue to perform these duties, for many, the original reason for building the railroad is gone. It often falls to rail enthusiasts to restore and preserve some of the world's most enchanting lines for posterity.

1. **The Durango & Silverton Narrow Gauge Railroad** (1881) in Colorado serviced gold and silver mines but is now a National Historic Landmark, running its original steam engines.   2. **The White Pass & Yukon Route** (1898) was Alaska's "railroad made of gold" during the Klondike Gold Rush. Closed in 1982, it is now run as a tourist attraction.
3. **The Ferrocarril Chihuahua-Pacifico, "El Chepe"** (1961), traverses Copper Canyon in Mexico. First planned in 1880, finances and the rugged landscape delayed construction.
4. **The Old Patagonian Express, "La Trochita"** (1935), in Argentina faced closure in 1992 but now runs more than 20 steam locomotives.   5. **The Furka Cogwheel Steam Railway** (1925) in Switzerland was abandoned when a mountain tunnel was built in 1982. Volunteers now operate trains up to Furka Station at 7,087 ft (2,160 m).   6. **Chemin de Fer de la Baie de la Somme** (1887) runs around part of the coast of northern France using vintage stock.   7. **The Historical Logging Switchback Railway** (1926) in Vychylovka, Slovenia, closed in 1971, but part of the line has been in operation for tourists since 1994.
8. **The Giant's Causeway & Bushmills Railway** (1883) in Ireland was a tramway powered by hydroelectricity. Closed in 1949, it reopened in 2002 using steam and diesel power.
9. **The Bluebell Railway** (1882) in England, formerly the Lewes and East Grinstead line, closed in 1958, reopening after two years as the world's first preserved standard-gauge passenger line.   10. **The Ventspils Narrow Gauge Railway** (1916) in Latvia was built by the German Army during World War I. Trains run today on a 1¼-mile (2-km) track.
11. **The Brocken Railway** (1898), or Brockenbahn, in Germany's Harz Mountains has run tourist steam trains to the Brocken mountain peak, at 3,743 ft (1,141 m), since 1992.
12. **The Puffing Billy Railway** (1900) once served farming and forestry near Melbourne, Australia, and is now preserved for tourists.   13. **The Darjeeling Himalayan Railway** (1881) is listed by UNESCO as one of the most outstanding examples of a hill railway in the world.

Clan Line & Belmond British Pullman

Built in 1948, Merchant Navy Class No. 35028 *Clan Line* operated as a mainline express passenger locomotive until 1967. Since *Clan Line* returned to service in 1974 it has been running special trips on Britain's main lines, and regularly hauls the *Belmond British Pullman* train. It has had three major overhauls in that time.

DESIGNED BY OLIVER BULLEID, *Clan Line* is one of 30 Merchant Navy Class 4-6-2 Pacific locomotives built from 1941 at the Southern Railway's Eastleigh Works. Each one was named after shipping companies that worked at the railroad's Southampton Docks. *Clan Line* worked on the Southern Region of the newly nationalized British Railways and in 1959 was rebuilt into its current form. In July 1967 it was sold to the Merchant Navy Locomotive Preservation Society, which later teamed *Clan Line* up with the *Belmond British Pullman* train.

The cars of the train once formed part of some of Britain's most famous services, such as the *Brighton Belle* and the *Queen of Scots*. After being withdrawn from service in the 1960s and 1970s, many fell into disrepair. In 1977, James B. Sherwood began buying and restoring the historic sleepers, saloons, and restaurant cars with the aim of reviving the legendary Orient Express; he acquired 35 cars in all.

FRONT VIEW

REAR VIEW

Coal space was extended in the 1995 overhaul to increase coal capacity

Tender has a sloping edge to the top to allow the crew better vision when moving backwards

Cab designed by Bulleid in consultation with crews

Nameplate mounted on the side

BR-type smoke deflectors added after the 1959 rebuild

Elegant locomotive
Clan Line is maintained in working order so that it can run on Britain's main lines. The entire class was rebuilt in the 1950s, so none survive in their as-built, "air-smoothed" condition.

SPECIFICATIONS FOR CLAN LINE			
Class	Merchant Navy	In-service period	1948 to present (*Clan Line*)
Wheel arrangement	4-6-2	Cylinders	3
Origin	UK	Boiler pressure	280 psi (19.69 kg/sq cm)
Designer/builder	Bulleid/Eastleigh Works	Drive wheel diameter	78 in (1,800 mm)
Number produced	30 Merchant Navy Class	Top speed	105 mph (167 km/h)

SPECIFICATION FOR CARRIAGES	
Origin	UK
In-service	various
Coaches	35
Passenger capacity	20-26 seats
Route	various

Hand rails on either side of the doors

Each car has two sets of 4-wheel trucks

Car names are displayed on the side

Pullman coat of arms displayed on the sides of each car

Palaces on wheels
The *Belmond British Pullman*'s cars run on the English leg of the famous Venice Simplon-Orient Express transcontinental train. They were restored to meet rigorous safety standards, yet maintain their stunning vintage features.

Traveling in style
Clan Line began its life hauling prestigious heavy boat trains with names such as *Golden Arrow* and *Night Ferry*. Nowadays *Clan Line* pulls the elegant carriages of the *Belmond British Pullman* train for Venice Simplon-Orient Express (VSOE), which offers a movable feast of fine dining and silver service.

The later BR logo
British Railways used this logo, known as the "ferret and dartboard crest," on steam locomotives from 1956.

LOCOMOTIVE EXTERIOR

The Merchant Navy Class locomotives were originally built with straight-sided, cylinder jacketing and an "air-smoothed," streamlined casing. The view from the cab was poor, and at high speed, the flat top generated a vacuum that drew exhaust steam down to obscure the driver's view. The engines were nicknamed "spam cans" because the shape of their casing resembled the tins of meat imported from the US at the time. The locomotives underwent a rebuild in the mid-50s emerging with a more conventional look, and the casing was replaced by conventional boiler cladding. *Clan Line* entered service painted in Southern Railway's Malachite Green livery, with British Railways' lettering.

1. Locomotive cab side number and power classification **2.** Headboard (when hauling the *Belmond British Pullman*) **3.** Electric headlight **4.** Draw hook and screw coupling **5.** Front buffer **6.** Right-hand compression valve and cylinder draincock pipes **7.** Cylinder draincock and pipes **8.** Walschaerts valve gear: slide bar, crosshead, piston, and radius arm **9.** Locomotive lubricators **10.** Whistle **11.** Nameplate **12.** Rear and middle drive wheels **13.** Boiler water injector (live steam monitor type) **14.** Water-injector delivery pipes to check valves **15.** Tender axle-box **16.** Drive wheel brake block

CAB INTERIOR

The Merchant Navy Class was reputed to have one of the safest cabs. The fire door contained air holes that minimized the risk of blowbacks, while levers allowed the engineer and fireman to operate the blowers without stepping in front of the firehole.

17. General arrangement of engineer's cab 18. Firebox door and back of boiler 19. Bottom of stack showing spark arrestor 20. Locomotive reverser winding handle 21. Vacuum brake ejector and steam brake control
22. Vacuum brake gauge (left) and steam chest gauge 23. Speedometer
24. Sand application valve, steam blower control behind (red handle)
25. Regulator handle 26. Automatic warning system (AWS) sunflower indicator 27. Boiler-gauge glass 28. Steam-supply control valve to air brake steam compressor (a modern addition) 29. Tender water-gauge indicator
30. Fireman's side control valves (left to right): damper controls, ashpan and tender sprinkler valves, steam and water-injector controls 31. Tender coal bunker space holding about 8 tons of coal

LUCILLE PULLMAN CAR EXTERIOR

Built in 1928, *Lucille* started out as a first-class parlor car for the *Queen of Scots* Pullman service, then ran in the *Bournemouth Belle*. *Lucille* joined the *British Pullman* train in 1986. Fully restored to its former glory, the sides of the car proudly display its name and the Pullman coat of arms against the gleaming umber and cream livery.

1. Pullman coat of arms transfers on side **2.** Car name painted in gold lettering **3.** Decorative gold embellishments on lower panels **4.** Decorative gold design on fascia embellishment **5.** Truck **6.** Owner's plaque **7.** Brass embarkation light above door **8.** Elaborately designed door handle **9.** Passenger door into car **10.** Specification plate attached to end of car

LUCILLE PULLMAN CAR INTERIOR

A prominent feature of this 1920s-themed, first-class dining car is the French-polished wood paneling of the walls and partitions. The distinctive marquetry on the side panels was manufactured by Albert Dunn in the 1920s and features Grecian urns on dyed green holly wood. The panel restoration was completed by the Dunn family using the original green veneer. Other period touches include Art Deco lampshades and brass fixtures. The windows beside each table are framed with curtains that add to the vintage feel. Luxurious upholstered dining chairs seat up to 24 passengers, who are served by staff dressed in period-style uniforms.

11. *Lucille* car ready for service **12.** Coat of arms and name plaque above doorway **13.** Gold lettering of car name above door **14.** Ceiling light with tulip glass shade **15.** Brass pull-down roof vent **16.** Public address speaker grille **17.** Decorative wooden marquetry panel featuring Grecian urn design **18.** Wall-mounted brass "torch lights" with tulip glass shade **19.** Embroidered motif on antimacassars **20.** Mirrors inlaid into wooden paneling **21.** Emergency stop chain **22.** Brass individual seat number **23.** Window latch **24.** Internal door handle **25.** Four-seat private coupe **26.** Washroom at end of car, accessed through corridor

CYGNUS PULLMAN CAR

Designed in 1938, *Cygnus*'s completion was delayed by World War II. Decorated with Australian walnut paneling, the car was reserved for use by traveling royalty and visiting heads of state. Along with *Perseus*, the car formed part of Winston Churchill's funeral train in 1965.

1. Single-seat layout of car 2. Pullman coat of arms on carpet at entrance to car 3. Washroom with mosaic floor depicting a swan 4. Washroom marble sink surround and wooden paneling 5. Decorative glass cathedral light window in the washroom

GWEN PULLMAN CAR

Originally on the iconic *Brighton Belle* service, *Gwen* is famous (along with sister car *Mona*) for conveying Queen Elizabeth (later the Queen Mother) to Brighton in 1948. After retirement, *Gwen* was bought by VSOE in 1988, restored, and returned to the rails as part of the *Belmond British Pullman*.

6. Coat of arms and name plaque above doorway 7. Decorative marquetry 8. Mirrored wooden panels divide length of car 9. View of *Gwen* car along central gangway 10. Kitchen situated at end of car, accessed through corridor

VERA PULLMAN CAR

Originally serving on the *Southern Belle* and *Brighton Belle*, *Vera* was badly damaged during the Blitz in 1940. It was returned to service seven years later when it was teamed up with *Audrey* as part of a five-car train. After retirement, *Vera* was used as a summer house in Suffolk until the Venice Simplon-Orient Express (VSOE) bought it in 1985. *Vera* is renowned for its gleaming interior, decorated with sumptuous panels of sandalwood and mahogany. The dining chairs are upholstered in a copy of the original 1930s fabric.

11. Coat of arms and name plaque over doorway 12. Wooden marquetry paneling 13. Overview of *Vera* car 14. Illuminated seat number 15. Etched-glass sunburst wall light 16. Detail of brass luggage rack 17. Four-seater coupe

ZENA PULLMAN CAR

This car was built in 1929 as a first-class parlor car for the Great Western Railway's *Ocean Liner* services to Plymouth. *Zena* had an illustrious career and even hosted the French President Auriol on a state visit in 1950. The beautifully restored sandalwood panels are inlaid with intricate motifs.

18. Coat of arms and name plaque above doorway 19. Art Deco motif on wooden paneling 20. Overview of *Zena* car ready for service

High Speed—The New Generation

By 2000, train speeds had increased since the first high-speed Shinkansen and TGV trains of the 1960s to 1980s. New lines were being designed specifically for trains that could operate at 200 mph (330 km/h), and a new generation of trains was being introduced in several countries. Plans for intercity Maglev (*Mag*netic *Lev*itation) routes were developed in both Germany and Japan, although, as yet, none has been built because the construction costs are too high. In the US, 150-mph (241-km/h) operation on sections of existing lines was introduced.

△ **Trenitalia ETR 500, 2000**

Wheel arrangement	13-car trains including two power cars
Power supply	3,000 V DC, 25 kV AC, overhead lines
Power rating	11,800 hp (8,800 kW)
Top speed	211 mph (340 km/h)

These trains were based on an earlier batch of ETR 500 trains built in the mid-1990s for operation on 3,000 V DC electrified lines of the Italian railroads (Trenitalia). The new high-speed lines connecting Naples and Rome, and Florence and Milan, which opened after 2000, needed trains able to work on 25 kV AC power. The latest ETR 500 can do this. The trains are limited to 186 mph (300 km/h) for current operation in Italy.

△ **DB ICE 3, 2000**

Wheel arrangement	8-coach, high-speed EMU
Power supply	15 kV AC, 16²⁄₃ Hz, 25 kV AC, 3,000 V DC, 1,500 kV DC, overhead lines
Power rating	10,724 hp (8,000 kW)
Top speed	205 mph (330 km/h)

Sixty-seven ICE 3 trains entered service from 2000, just before the new 205-mph (330-km/h) high-speed line connecting Cologne with Frankfurt airport opened in 2002. All had eight cars and featured panorama lounges at either end, where passengers could see the driver and the line ahead through a glass screen. Seventeen of the trains were four-voltage international sets, four of which were bought by Dutch Railways.

▷ **SMT/Transrapid, 2004**

Wheel arrangement	Maglev (no wheels)
Power supply	electromagnetic suspension
Power rating	unknown
Top speed	268 mph (431 km/h)

The world's first commercial Maglev system was built at Birmingham Airport, England, in 1984. Development of Maglev systems was led by Japanese and German companies in the 1990s, and the world's only high-speed system opened in China in 2004, with a 19-mile (31-km) route connecting Shanghai city to Pudong International Airport using German-built trains.

◁ **Chinese Railways CRH₂A, 2007**

Wheel arrangement	8-coach high-speed EMU
Power supply	25 kV AC, overhead lines
Power rating	6,434 hp (4,800 kW)
Top speed	155 mph (250 km/h)

The Chinese government ordered 60 CRH₂A trains from Kawasaki of Japan, working with China Southern Rolling Stock Corp. (CSR), in 2004. The train is based on the E2 Shinkansen operated by Japan Railways (JR) East. The first three were built in Japan; the remainder were assembled at CSR Sifang. CSR has built several more versions since 2008, including 16-car sleepers.

△ Amtrak Acela, 2000

Wheel arrangement 2 x Bo-Bo power cars plus 6 passenger cars

Power supply 11 kV AC 25 Hz, 11 kV AC 60 Hz, and 25 kV AC 60 Hz, overhead lines

Power rating 12,332 hp (9,200 kW)

Top speed 165 mph (266 km/h)

Amtrak ordered the new Acela design following trials of several European high-speed trains in the US in the 1990s. Built to unique US standards for crashworthiness, the trains can tilt, allowing higher speed on curves. Current maximum speed is 150 mph (241 km/h) in service, but plans exist for 160-mph (258-km/h) operation on some sections of the Washington, DC–Boston route in the future.

▽ RZD *Sapsan*, 2009

Wheel arrangement 10-coach, high-speed EMU

Power supply 25 kV AC, 3,000 V DC, overhead lines

Power rating 10,728 hp (8,000 kW)

Top speed 155 mph (250 km/h)

German train manufacturer Siemens built eight 10-car broad-gauge (5-ft/1.52-m) versions of its Velaro high-speed train for Russia in 2009–11. Russian Railways (RZD) operate the trains branded *Sapsan* (peregrine falcon)–the fastest bird– between Moscow and St. Petersburg/Nizhny Novgorod. Eight more trains are due to enter service in 2014-15.

△ JR N700 Shinkansen, 2007

Wheel arrangement 16-coach, high-speed EMU

Power supply 25 kV AC, overhead lines

Power rating 22,905 hp (17,08 kW)

Top speed 186 mph (300 km/h)

The N700 Shinkansen can accelerate faster than any of the trains it replaced on the Tōkaidō Shinkansen line between Tokyo and Hakata. Built in either 8- or 16-car train sets, most N700s were in service by 2012, and 149 trains will have been delivered by the time production ends in 2016.

TALKING POINT

Meals on the Move

"Bento" is the Japanese name for a carefully crafted packed meal in a single, often disposable, container. Bento boxes were historically made from wood or metal, but are now found in a variety of materials and novelty shapes. A wide range of bento box train meals, known as "ekiben," are sold at kiosks in train stations all over Japan to take on board the train.

Novelty boxes
Ekiben packed in boxes shaped like Japanese trains have become collector's items. This one is modeled on the N700 Shinkansen.

Spectacular Stations

Train stations have become sites of some of the world's most exquisite architecture and design. Whether the look is contemporary or classic, the architecture of the most celebrated stations successfully combines form and function to make an indelible impression on every traveler who passes through.

1. The Tsuzumi Gate of Kanazawa Station in Tokyo (2005) combines traditional Shinto temple designs with the strings of a Japanese drum.
2. Danggogae Station in Seoul (1993) is the north terminus of Line 4 of the South Korean capital's subway. **3. Financial Centre metro station in Dubai** (2009) was designed with a shell-like structure that recalls the city's early prosperity from pearl diving. **4. Komsomolskaya Station in Moscow** (1952) is located on the metro's Koltsevaya Line. Its grand architecture features chandelier lighting, baroque details, and mosaics of historic Russian scenes.
5. Haydarpasa Terminus in Istanbul (1908) has neoclassical styling, is surrounded on three sides by water, and is Eastern Europe's busiest station.
6. Tanggula Station in Tibet (2006) is the highest train station in the world at 16,627 ft (5,068 m) above sea level. **7. Chhatrapati Shivaji Terminus in Mumbai** (1888) has Gothic-style architecture and has been recognized by UNESCO as a world heritage site. **8. Berlin Hauptbahnhof** (2006) is a glass and steel structure with platforms on two levels, serving 350,000 passengers daily. **9. St. Pancras International in London** (1868) reopened in 2007 following extensive renovation and expansion, but retained the original Victorian roof, to dramatic effect. **10. Liège-Guillemins Station in Liège** (2009) has no outer walls, but a glass and steel canopy that covers all five platforms. The station serves as a transportation hub for high-speed rail links across Europe. **11. Grand Central Terminal in New York** (1913) has 44 platforms—more than any other station in the world. The Beaux-Arts architecture features Botticino marble staircases and an astronomical ceiling.
12. Union Station in Los Angeles (1939) has the appearance of a church from the outside, but is one of the busiest stations on the West Coast, serving more than 60,000 passengers daily. **13. Ushuaia Station in Tierra Del Fuego** (1910) was originally used to transport prisoners to an Argentinian penal colony. It closed in 1947, but reopened in 1994 after extensive renovation; the station has since become a popular tourist attraction.

Faster and Faster

Many existing high-speed train fleets have been expanded and, as new lines have opened, increasing numbers of fast international services have become possible, connecting France, Germany, Spain, and Switzerland in particular. In France, a specially modified test train built by Alstom, the TGV V150, achieved a new world speed record of 357¼mph (574.8 km/h) in 2007. In Japan, the main railroads had been privatized by 2006, and in 2012 the world's first "start-up" private high-speed operator was NTV in Italy, which also began its services with a brand-new train design. However, government-owned operators continue to dominate.

△ NTV AGV ETR 575, 2012

Wheel arrangement	11-coach, articulated, high-speed EMU
Power supply	3,000 V DC, 25 kV AC, overhead lines
Power rating	10,054 hp (7,500 kW)
Top speed	186 mph (300 km/h)

Italian private rail operator Nuovo Trasporto Viaggiatori (New Passenger Transport), started high-speed services from Naples to Rome and Turin in 2012 with a fleet of 25 Alstom-built AGV (Automotrice à Grande Vitesse) high-speed trains. The trains are equipped with three different classes of passenger accommodations.

△ VT Class 390 Pendolino, 2002

Wheel arrangement	9- or 11-coach, high-speed tilting EMU
Power supply	25 kV AC, overhead lines
Power rating	7,976 hp (5,950 kW)
Top speed	124 mph (200 km/h)

The Virgin Trains's tilting, high-speed Pendolino trains have speeded up journeys on the UK's West Coast Main Line from London to Birmingham, Manchester, and Glasgow since 2002. By leaning on curves, the train can go faster than conventional ones, reducing trip times and increasing track capacity.

TALKING POINT

Steam Train Revival

The UK has led the world in preserving mainline steam locomotives, with many restored to working order since the 1950s. The success in preserving different types has led to groups of volunteers trying to recreate engine classes that never survived. All 49 Peppercorn Class A1 locomotives were scrapped in the 1960s, so in 1990 a group decided to build another—from scratch. Nineteen years later, the new engine, based on the original design but with some modern features, started operations. It now works charter trains all over the country. Several other similar projects are now underway in the UK.

Peppercorn Class A1 No. 60163 _Tornado_, 2008
This is the first mainline steam engine built in the UK since 1960.

▽ LSER Class 395 Javelin, 2009

Wheel arrangement	6-car EMU
Power supply	25 kV AC overhead lines and 750 V DC third rail
Power rating	4,504 hp (3,360 kW)
Top speed	140 mph (225 km/h)

Built by Hitachi in Kasado, Japan, using Shinkansen technology, the Javelin trains have been in service since 2009 serving the London & South Eastern Railway on the UK's domestic high-speed line HS1. They have reduced travel times significantly (in some cases by as much as 50 percent) between cities in Kent and London.

△ PKP IC Class ED250, 2014

Wheel arrangement 7-coach, high-speed EMU

Power supply 15 kV AC, 16²/₃ Hz, 25 kV AC, 3,000 V DC, 1,500 kV DC, overhead lines

Power rating 7,376 hp (5,500 kW)

Top speed 154 mph (249 km/h)

Poland's long-distance railroad company Polskie Koleje Panstwowe Intercity (PKP IC) ordered 20 Class ED250 trains from Alstom. Based on the "New Pendolino" design, first built for China and Italy, but without the tilt equipment, the trains began to replace older locomotive-operated trains on the Gdynia/Gdansk–Warsaw–Krakow/Katowice route from late 2014.

△ SNCF TGV POS, 2006

Wheel arrangement 10-car train including 2 power cars

Power supply 15 kV AC, 16²/₃ Hz, 25 kV AC, 1,500 kV DC, overhead lines

Power rating 12,440 hp (9,280 kW)

Top speed 199 mph (320 km/h)

Using the new LGV Est high-speed line that connects the Alsace region with Paris, the French national railroads' (SCNF) TGV POS (Paris-Ostfrankreich–Süddeutschland, or Paris-Eastern France–Southern Germany) trains started operating in 2006. The TGV POS trains also work express services to Munich and Frankfurt, Germany, plus Geneva and Zurich, Switzerland.

◁ SNCF TGV Euroduplex, 2012

Wheel arrangement 10-car train including 2 power cars

Power supply 15 kV AC, 16²/₃ Hz, 25 kV AC, 1,500 kV DC, overhead lines

Power rating 12,440 hp (9,280 kW)

Top speed 199 mph (320 km/h)

The third-generation "Duplex" (double-deck) TGV train was built for the French national railroad SNCF; 95 have been ordered, with deliveries planned until 2017. This is the only double-deck high-speed train capable of operating across several different European rail networks and is used for services between France and Germany and France and Spain.

▷ SNCF TGV V150, 2007

Wheel arrangement 5-car TGV train

Power supply 31 kV AC, overhead lines

Power rating 26,273 hp (19,600 kW)

Top speed 357¹/₄ mph (574.8 km/h)

This test train used two new TGV POS power cars and three special cars with powered trucks, all vehicles having specially made, bigger wheels. For the test run on the LGV Est Européenne line, the overhead line voltage was increased for more power. The record of 357¹/₄ mph (574.8 km/h)—6 miles (9.65 km) per minute—more than achieved the target set.

southeastern. highspeed

39176

Javelin No. 395 017

Built by Japanese manufacturer Hitachi, the Class 395 Javelin is based on technology used in Shinkansen trains. A multiple-unit train set with a powercar at each end, the Javelin can be powered by overhead wires, but uses a third-rail electricity pickup when operating on conventional lines in southeast England. Javelin trains have reduced some journey times by as much as 50 percent, and they provided a key service during London's 2012 Summer Olympic Games.

THE HIGH-SPEED LINE from London to the Channel Tunnel, known as High Speed 1 (HS1), was finished in November 2007. Providing international services via the tunnel, it also allows domestic high-speed Class 395 Javelins to serve southeast England. Although slower than the 189-mph (304-km/h) Eurostar trains with which it shares the HS1, the lighter, shorter Javelin accelerates faster.

The Javelin has four safety systems. Two French systems are used on the HS1; Transmission Voie Machine (TVM430) transmits signaling information through the track to displays in the driver's cab, and Contrôle Vitesse par Balise (KVB) monitors and controls the train's speed from St. Pancras International station, London. On conventional British routes, the Automatic Warning System (AWS) and Train Protection and Warning System (TPWS) work together to alert the driver to signals, and stop the train if it passes a danger signal.

FRONT VIEW (DPT2)

FRONT VIEW (DPT1) - EXTRA YELLOW PANELS INDICATE UNIVERSAL ACCESS

SPECIFICATIONS

Class	395	**In-service period**	2009–present
Wheel arrangement	6-car EMU	**Railroad**	Southeastern
Origin	Japan	**Power supply**	25 kV AC overhead wires and 750 V DC third rail
Designer/builder	Hitachi, at Kasado	**Power rating**	4,506 hp (3,360 kW)
Number produced	29	**Top speed**	140 mph (224 km/h)

High-speed designer
German designer Alexander Neumeister created the look of the Javelin, which bears a resemblance to his German ICE 3 and several Japanese Shinkansen variants. Neumeister has also designed Chinese and Russian high-speed trains.

Sliding doors are automated

Safety line warns rail staff of the presence of overhead power cables

Electrical equipment including heating and air-conditioning units

Name honors one of 24 British athletes from the 2012 London Olympics

Third-rail shoe picks up electrical power from third rail

southeastern. highspeed

Britain's Fastest
Dame Sarah Storey

southeasternrailway.co.uk

395 017

Javelin

Serving the Olympic Games
Ferrying millions of visitors to the Olympic Park via HS1 and the station at Stratford International, the 2012 Olympic shuttle was known as the Javelin, and the name endures today.

EXTERIOR

The aluminum car bodies are adorned in a dark blue livery that is unique to the Javelin trains. Each car has two wide single sliding doors on each side, painted in lighter colors to be easily identified by passengers. The driving cars at each end hold the pantographs and third-rail collector shoes that pick up electricity to power the train.

1. Olympian signature on side of front vehicle   2. Headlight (above) and taillight (below)   3. Coupler and horn inside open nose cone   4. External emergency door release (access handle)   5. Driving cab door handle   6. Driver's cab door   7. Southeastern logo on side   8. Rheostatic brake resistor mounted on roof   9. Pantograph assembly   10. Vacuum circuit breaker (VCB)   11. Third-rail shoe fuse   12. Axle end earth on rear vehicle   13. Underframe view of motor truck

Sarah Storey

Britain's Fastest

Dame Sarah Storey
22 Time Paralympic Medallist

southeasternrailway.co.uk

CAB INTERIOR

The cab is typical of those found in modern high-speed trains, with a single driver's chair positioned centrally, facing the control desk. To the right are CCTV displays from the passenger cars and to the left is the Train Management System (TMS) which, among other functions, allows the driver to switch between third-rail and overhead power.

14. Driver's seat and controls 15. Train management system (TMS) 16. Combined power/brake controller (CPBC) 17. CCTV panel position in driving cab 18. Emergency brake push button 19. Master key 20. Secondary seat in driver's cab 21. Short-circuiting bar and red flag, for use in emergencies 22. Miniature circuit breaker (MCB) panel 23. Switch panel on cab back wall 24. Onboard manager's door control panel in driving cab

CAR INTERIORS

The Class 395 is designed for commuter trips that typically last no more than an hour. The train does not offer first-class accommodations, but its passengers pay higher fares than they would on the slower conventional trains. The gangway runs through all six cars and connects them, but if two trains are joined together, it is not possible to get from one six-car section to the other.

The interior of the carriages is blue and gray, and complements the dark blue external livery. Each train contains 340 2+2 seats, with the majority of them facing in the same direction. In addition, there are 12 tip-up seats in the door vestibule areas. Two lavatories per train are provided, one of which is designed for universal access. Passenger information interfaces on the train include digital displays and a PA system.

25. Overview of car **26.** Seats with table **27.** Luggage rack above seats **28.** Adjustable arm rest **29.** Tray table on seat back **30.** Interior luggage area **31.** Gangway door between cars **32.** Gangway door open button **33.** Passenger information system (PIS) display panel **34.** Disabled-access lavatory **35.** Power outlet sign above seats **36.** Emergency alarm sign **37.** Handhold on seat for passengers walking in the aisle **38.** Interior of disabled-access car **39.** Passenger door open/close buttons

Dubai Metro

Worsening traffic congestion and a growing population—expected to reach 3 million by 2017—persuaded Dubai's leaders to build the United Arab Emirates' first subway. In May 2005 a US$3.4 billion contract was awarded to Dubai Rail Link (DURL), a consortium of companies from France, Japan, Turkey, and the United States. Two routes were planned: the Red Line and the shorter Green Line. Work began in 2006 on the Red Line, with stations serving central Dubai.

The opening of the first section on September 9, 2009 attracted more than 110,000 people. The five-car trains could carry up to 643 passengers and provided three classes of travel, including a section for women and children, but the most impressive feature was that the trains were fully automated. In fact, this was the inaugural stage of what—at 47 miles (75 km)—was declared in 2012 to be the world's longest driverless rail network.

PLANS FOR EXPANSION

The Red Line opened fully in April 2010, by which time the inaugural section had already carried more than 11 million passengers. The second route, the Green Line, was opened in September 2011. The Metro's success has led to plans to extend the existing lines and create three additional routes which, by 2030, would enlarge the Dubai Metro to 262 miles (421 km) of track, servicing a total of 197 stations.

The driverless trains on the Red Line Metro in Dubai pass over viaducts above the streets as well as underground. Power is drawn from a third rail.

Into the Future

Investment in railroads around the world is growing, driven by rising passenger numbers as large cities continue to expand, increasing road congestion, and the need to reduce CO_2 emissions. Older-style trains that use locomotives and separate coaches are being replaced by modern self-powered multiple units. Rail operators, both passenger and freight, are also seeking to reduce maintenance and energy costs; some modern trains are designed to recycle electricity while braking.

◁ **Bombardier Omneo Régio2N, 2010**

Wheel arrangement	6- to 10-car articulated EMU
Power supply	25 kV AC, 1,500 kV DC, overhead lines
Power rating	4,291 hp (3,200 kW)
Top speed	99 mph (160 km/h)

The Omneo is the world's first articulated, double-deck EMU with single-deck driving coaches at each end, and double-deck, articulated intermediate coaches sandwiched between short, single-deck door sections. The trains can be supplied in lengths ranging from 6 to 10 cars (266–443 ft/81–135 m). The French national railroad (SNCF) has approved a €7 billion ($9.7 billion) contract for up to 860 trains for delivery until 2025.

△ **Bombardier Zefiro 380, 2012**

Wheel arrangement	8-car EMU
Power supply	25 kV AC, overhead lines
Power rating	13,454 hp (10,037 kW)
Top speed	236 mph (380 km/h)

The latest version of the Bombardier-designed Zefiro high-speed train operates at up to 236 mph (380 km/h). Chinese Railways have ordered 70 (two were delivered in 2012). In Europe, 50 224-mph (360-km/h) versions built for Italian operator Trenitalia, entered service in 2014.

Local Transportation Developments

The demand for urban transportation has grown significantly in the last 30 years—whether metros under city streets or light-rail systems that run on roads alongside other vehicles. The strongest growth is in Asia and the Middle East, where new systems have been built since 1990. For established networks, the challenge is to create more capacity through better performance and smart control systems on networks that are more than 100 years old—for example, in London and Paris.

▷ **Vossloh Wuppertal Schwebebahn train, 2015**

Wheel arrangement	3-section, articulated vehicle
Power supply	750 V DC, third rail adjacent to single running rail
Power rating	322 hp (240 kW)
Top speed	approx. 37 mph (60 km/h)

Germany's Wuppertal Schwebebahn is a suspended railroad built largely above the River Wupper on massive iron supports. First opened in 1901, it is now a protected national monument, but is still used daily by thousands of commuters. Vossloh will supply 31 new trains from 2015 as part of a comprehensive modernization plan.

◁ **Amtrak Siemens American Cities Sprinter ACS-64, 2014**

Wheel arrangement Bo-Bo

Power supply 25 kV, 12.5 kV, and 12 kV AC, overhead lines

Power rating 8,579 hp (6,400 kW)

Top speed 125 mph (201 km/h)

Siemens is building 70 ACS-64s at its factory in Sacramento, CA. Amtrak, which introduced the first ACS-64 in 2014, will use them to replace all its existing electrics on the Washington, DC–New York–Boston Northeast Corridor route.

△ **VMS Chemnitz tram-train, 2015**

Wheel arrangement 3-section articulated LRV

Power supply 600 V and 750 V DC, overhead lines plus diesel engines

Power rating electric: 777 hp (580 kW); diesel: 1,046 hp (780 kW)

Top speed 62 mph (100 km/h)

Tram-trains that enable travel to downtown stations from regional rail lines are now in use in many EU countries. In Germany, some use diesel engines on non-electrified rail lines. Chemnitz tram-trains will use this technology from 2015.

TECHNOLOGY

Cargo Efficiency

The major Class 1 railroads in North America have increased operational efficiency and productivity significantly since the 1980s. Pulling longer, heavier trains with powerful modern locomotives has reduced operating costs per cargo container, making rail much cheaper than road. Double-stacked containers are used in North America, Australia, and India. The Brazilian mining company Vale runs the 554-mile (892-km) Carajás Railroad with the world's heaviest trains: 330-car, 46,600-ton (42,300-metric-ton) iron-ore trains run up to 24 times a day to the port at Ponta da Madeira.

BNSF freight train, Cajon Pass, California With two modern GE Evolution Series ES44DC engines at each end, this train can be up to $2^2/_3$ miles (4.3 km) long. On steep gradients, the engines slow down descending trains, as well as pull them up the inclines.

△ **Siemens Vectron, 2013**

Wheel arrangement Bo-Bo

Power supply 3 kV DC, overhead lines

Power rating 6,970 hp (5,200 kW)

Top speed 99 mph (160 km/h)

Siemens developed the Vectron family of locomotives to replace its previous Eurosprinter model. The first major order received was for 23 Vectron DC electric locomotives from Polish rail freight operator DB Schenker Rail Polska; the first of these entered service in 2013. Subsequent orders for locomotives for use in several countries have been obtained, including a broad-gauge version for Finland.

△ **Siemens ICx, 2017**

Wheel arrangement 7- or 12-coach, high-speed EMU

Power supply 15 kV AC, 16$^2/_3$ Hz

Power rating 13,271 hp (9,900 kW)

Top speed 155 mph (250 km/h)

The "ICx" trains will replace Germany's existing, long-distance, locomotive-operated trains, and later the first two types of ICE trains. Due for delivery from 2017 are 85 12- and 45 slower 7-coach trains using 92-ft- (28-m-) long coaches configured as distributed-power EMUs with more seats and space than those they replace.

△ **Calgary Transit C-train System Siemens S200, 2015**

Wheel arrangement 2-car, articulated LRV

Power supply 600 V DC, overhead lines

Power rating 777 hp (580 kW)

Top speed 65 mph (105 km/h)

The Canadian city of Calgary opened its first light-rail line in 1981. Since then, the network has expanded and carries 290,000 people daily. To increase capacity and to retire some of the original light-rail vehicles (LRVs), 60 new S200 LRVs are on order for delivery in 2015–16. Calgary Transit expects to increase its fleet from under 200 to 390 over the next 30 years.

△ **London Underground Siemens Inspiro metro concept**

Wheel arrangement 6-car, metro EMU

Power supply 630 V DC, third and fourth rail

Power rating 1,341 hp (1,000 kW)

Top speed 56 mph (90 km/h)

London Underground has seen significant growth in passengers. The "New Tube for London" program is planning 250 new underground trains, possibly automatic and driverless, to enter service between 2020 and 2035. Three companies are designing trains; shown here is Siemens's proposal.

HOW RAILROADS WORK
ENGINES
AND TRACKS

How Tracks Work

Wooden rails were used for the pony-drawn wagonways of the 17th century, but iron was required to support the steam engines of the 19th century. The cast-iron rails of the first railroads were succeeded by sturdier wrought-iron rails in the 1820s, before steel—stronger still—came into use. Steel rails were first laid at Derby station in England in 1857. The rails, ties, and ballast of a finished rail line have come to be known as the "permanent way," a term that dates back to the earliest

days of railroad construction, when a temporary track was laid first in order to transport materials to where the line was being constructed. The temporary track was replaced by the permanent way once the substructure was largely completed. The "gauge"—the distance between the rails—and the alignment of the rails are constantly monitored during construction to ensure that they remain uniform throughout the straight sections and curves in the track.

TRACK FORMATION

The substructure of a track is called the formation. Since a consistent grade (gradient) is required for trains to run smoothly, the ground is first prepared to form the subgrade. The subgrade might also be covered by a layer of sand or stone called a blanket before it is overlaid with ballast. Ties are bedded into the ballast to support the rails. Crushed-stone ballast is still the most common foundation and allows for good drainage.

SIDE VIEW

Ballastless track
Although expensive to install, ballastless track using a concrete roadway or precast concrete members saves maintenance costs.

Ladder track
Ladder track uses ties running in the same direction as the rails, with cross-member "rungs" to maintain the gauge.

TRACK STRUCTURE

Most modern railroad tracks consist of flat-bottom steel rails attached to wooden or concrete ties. Flat-bottom steel rails are more stable, easier to lay, and do not suffer from wear in the same way as the old cast-iron or wrought-iron rails. Bullhead rails are the same shape on the top and on the bottom so that they can be turned over and reused when the head becomes worn.

FLAT-BOTTOM RAIL   **BULLHEAD RAIL**

TRACK GAUGE

The gauge of a railroad's tracks is defined as the distance between the rails, measured from the inside of the rail—except in Italy, where it can be the distance between the center of the rails (see below). The first railroad builders chose whatever gauge they felt was appropriate for their line; a wider gauge was thought to give greater stability for a train at high speed or in strong crosswinds, while a narrow gauge took up less space and was usually cheaper to build. When lines grew into networks, some form of standardization became essential.

Two-foot
(2 ft/0.61 m) Category includes 1 ft 11 5/8 in (0.6 m). Industrial and military use, worldwide.

Russian
(4 ft 11 27/32 in/1.52 m) Former Soviet Union. Second most used after standard gauge.

Italian meter
(3 ft 1 3/8 in/0.95 m) Italy, Sardinia, and Sicily. Meter measured from center of rail.

Irish
(5 ft 3 in/1.6 m) Used in Ireland, Brazil, Switzerland, Germany, Australia, New Zealand.

Meter
(3 ft 3 3/8 in/1 m) Mountain rail and tramways worldwide, some light metros.

Iberian
(5 ft 5 21/32 in/1.67 m) Allows compatibility between Spanish and Portuguese gauge.

Cape
(3 ft 6 in/1.067 m) Adopted in 1873 in Cape Colony (South Africa). Africa, Japan, worldwide.

Indian
(5 ft 6 in/1.676 m) India and Pakistan. Called Portland in US, Provincial in Canada.

Scotch
(4 ft 6 in/1.372 m) Early Scottish railroads including Monkland and Kirkintilloch (first public).

Brunel
(7 ft 1/4 in/2.14 m) Used on Isambard Brunel's GWR from 1838 to 1892.

Standard
(4 ft 8 1/2 in/1.435 m) Used on 60 percent of the world's railroads, including US and UK.

Breitspurbahn
(9 ft 10 in/3 m) Proposed for Hitler's Third Reich supertrain but never built.

How Wheels Work

The wheels of a train are designed to enable it to follow curves in the track. Each wheel tapers from the inside outward and has a projecting flange on the outer edge. The flange is to prevent the wheels from derailing and normally never comes into contact with the track, the weight of the train being borne by the conical surface. These sloping edges allow the wheels to slide across the tops (heads) of the rails. The wheels on the outside of a curve have farther to travel and use the larger radius close to the flange, while the wheels on the inside use the shorter radius closest to their center.

The flanged wheel
The flanged wheel was invented by English engineer William Jessop in 1789 to provide a better grip on railed track; this helped prevent derailments.

STEAM LOCOMOTIVE WHEEL CONFIGURATION

As steam locomotives grew bigger and heavier, they gained more wheels, spreading their weight more evenly and giving better traction. To describe the wheel arrangement, mechanical engineer Frederick M. Whyte came up with a numbering system in 1900. A locomotive with four leading wheels, four powered wheels, and two trailing wheels was a 4-4-2. Articulated locomotives, designed to tackle bends more easily, needed longer numbers, but the codes retained their simple logic. The Whyte system is used in the US and the UK for steam engines, although different systems are used elsewhere, and for other types of locomotive. Some configurations also had names.

WHEELS	TYPE	NAME
	0-2-2	Northumbrian
	2-2-0	Planet
	2-2-2	Jenny Lind or Patentee
	4-2-0	One-Armed Billy
	0-4-0	Four-Wheeler
	4-4-0	American or Eight-Wheeler
	4-4-2	Atlantic
	2-6-0	Mogul
	2-6-2	Prairie
	4-6-0	Ten-Wheeler or Grange
	4-6-2	Pacific
	4-6-4	Baltic or Hudson
	2-8-0	Consolidation
	2-8-2	Mikado, Mike, or MacArthur
	2-8-4	Berkshire
	4-8-4	Northern
	2-10-4	Texas or Selkirk
	0-4-4-0	*none*
	2-6-6-2	*none*
	2-6-6-6	Challenger
	4-6-6-4	Blue Ridge or Allegheny
	2-8-8-4	Yellowstone
	4-6-4+4-6-4	Double Baltic

BRAKES

The first train brakes worked like the brakes on a horse-drawn cart, with levers moving a brake shoe to press a wooden block against the wheel tread. This was not very efficient and caused wear. Modern trains use disc brakes like those used on cars. The discs are attached to the axles, and calipers equipped with composite brake blocks "pinch" the discs to slow the train.

RIM BRAKE   **DISC BRAKE**

AIR BRAKE

During the 1870s two different types of brake systems were tried a vacuum system and an air brake system. Air brakes were shown to bring a 15-car train traveling at 50 mph (80 km/h) to a halt in half the time taken by vacuum brakes. The braking distance for the Westinghouse air brake was 777 ft (237 m), while a vacuum brake took 1,477 ft (450 m). Air or pneumatic braking is the standard system used today by the world's railroads.

Air brake application
A pump compresses air for use in the system. The driver controls the air with a triple valve. When this is applied, compressed air is released into the brake pipe and air pressure forces the piston to move against a spring in the brake cylinder, causing the brake blocks to make contact with the wheels.

Air brake release
When the driver releases the brake valve, air leaves the brake pipes. As air escapes from the exhaust, a spring in the brake cylinder pushes the piston back, causing the brake blocks to disengage from the wheels. The auxiliary air reservoir, meanwhile, refills.

How Signals Work

In the earliest days of the railroads, there were few trains and no real need for signaling systems. Trains ran up and down single tracks, and schedules kept them far enough apart to avoid accidents. As rail networks became more extensive, with rail traffic traveling from far and wide, timetables based on local time (most nations did not have standard time until the late 19th century) caused huge confusion, and signaling became essential to prevent collisions. By the 1830s the hand and lamp signals used by rail staff were being imitated by more visible mechanical trackside signals, although in some countries it would take almost a century for the style of these signals to be standardized across different networks.

Signal lamp
Lamps used by train guards, brake men, or station staff had different lenses that could shine a red, green, or clear white light.

EARLY SIGNAL SYSTEMS

The first trackside signals came in a variety of forms but, like the signaling lamps employed before them, used the color red to mean "stop." Long recognized as the international color for danger, red was an obvious choice. Green for "go" had also been used in lamps and was chosen because it couldn't easily be mistaken for red, or for a non-signal light that a locomotive engineer might happen to see. Yellow lights, the color adopted because it was distinct from the other two, were later introduced to advise caution. The trackside semaphore style of signal became the most widespread type and is still in use.

BALL TOKENS USED ON INDIAN RAILWAYS

Signaling tokens
Tokens were used on single lines to ensure only one train could enter each "block" (section) of the line at a time. The crew picked up a token from a signalman when it was safe to enter the block, handing it to another signalman at the end of the block. Automated systems were later developed with tokens dispensed and recovered by machines.

Ball signal
The most common signal on the early US railroads, the ball signal gave rise to the term "highball". When it was raised, it was safe to proceed. This was later reversed.

Semaphore
Widespread after the 1850s, and still in use today, semaphore arms signaled "danger" when in the horizontal position and "all clear" when angled either up or down.

Wood's crossbar signal
Crossbar signals, in use from the 1830s, indicated on/off (stop/go) with a revolving wooden board. When the crossbar was swung parallel to the line it signaled clear.

Revolving disk signal
The disk revolved vertically to signal stop and go, much like semaphore signals. In keeping with most signals of the time, the disk was made of wood and painted red.

Double disk signal
Like the crossbar, the double disk rotated on a wooden or steel signal post. Both were short-lived, however, as the "clear" signal was hard for train drivers to see.

ELECTRIC LIGHT SIGNALS

The use of electric signal lights instead of oil lamps started to become common in the 1920s, although early electric lights were still not powerful enough for drivers to see them clearly from a safe distance in daylight. Semaphore signals were far clearer and easier to spot. It was not until 1944 that modern lenses improved the visibility of electric signals sufficiently to allow them to replace semaphore signals fully. Railway light signals have red lights at the bottom so that they are in the engineer's line of sight; road traffic lights have red lights at the top so that drivers can see them above other cars.

Unlit glass (yellow)

Unlit glass (green)

Lit glass (yellow)

Unlit glass (red)

FRONT VIEW

Lamp shield

Lifting lug

Clip

Base

SIDE VIEW

Stopping the train
Signal lights control their own block of track and are designed to give a driver all the warning he needs to slow down before reaching a hazard. Modern rail networks also have safety systems in place to apply a locomotive's brakes automatically if it passes a danger signal.

Green "all clear" light
tells Train A to proceed into next block of track

Red "stop" light tells following train not to enter this block of track

Train A

SEMAPHORE SIGNAL

Semaphore signal systems began to be introduced in the 1840s and consisted of two pivoting arms or "blades" and a "spectacle" holding two colored glass lenses. As the arm moved, the lenses moved in front of a light source, initially an oil lamp, to allow the signals to be seen at night. On the top arm the lenses were red and blue, the blue combining with the yellowish flame of the oil lamp to make green. On the bottom arm the lenses were yellow and blue. Once electric lights were used, the blue lenses were changed to green.

Red, square-ended arm is horizontal, meaning "stop"

Yellow distant warning arm indicates "proceed with caution"

Yellow warning arm is raised, meaning "all clear"

Stop
When the upper arm is horizontal, it means "stop." The lower arm is a "distant" warning, telling the driver that the train may have to stop at the next signal. Both arms horizontal means "stop."

Proceed with caution
With the upper "stop" arm raised and its light green, the train can proceed, but the lower distant warning arm is still telling the driver to be cautious because the next signal may require the train to stop.

All clear
When both upper and lower arms are raised and both lights show green, it means that the line ahead is clear. The driver can proceed safely at normal speed to the next signal.

Telegraph

The telegraph transformed railroad signaling, allowing messages to be sent ahead of trains for the first time. The most successful telegraph system was invented by Samuel Morse in 1835.
  Morse's first apparatus used a pendulum device, but his partner, Alfred Vail, suggested using a lever and armature to print a code of dots and dashes—the precursor to Morse code. The system was patented in 1840 and adopted for railroad signaling and general use. Thousands of miles of telegraph lines were strung alongside train tracks, connecting the East and West coasts in October 1861.

Morse key sounder, 1875 The Morse telegraph used a single electric current switched on or off to send a series of dots and dashes.

SWITCHES

A switch mechanism is a key component of any modern railroad. Invented by English engineer Charles Fox in 1832, the simplest system uses a pull rod that is activated by a lever. This adjusts movable track sections (switches), to direct a train onto a curve, taking it away from the main line. Changing the switches was a task performed by a signalman, but most switches are now operated electronically.

Main line open

No signal for branch line

Switch stand or lever

Pull rod extended

Branch line closed

Switch B closed

Switch A open

Switching
Retracting the pull rod closes Switch A and opens Switch B. The train wheels then follow the tracks around onto the branch line, which will then be open, while the main line will be closed.

Modern switches
Modern switch systems are now controlled electronically, allowing for more complex traffic operations at busy sections of connecting track.

Green "all clear" light tells Train A to proceed into next block of track

Green "all clear" light tells Train A to proceed into next block of track

Two yellow "preliminary caution" lights instruct Train A that it must expect to stop two signals' ahead

Yellow "caution" light instructs Train A that it must expect to stop at next signal

Red "stop" light instructs Train A not to enter this block of track

Green "all clear" light instructs Train B to proceed into the next block of track

Train B

Radstock North Signal Tower

Before the days of automated signaling centers, signalmen managed the movement of trains from local signal towers. The Radstock North tower once controlled trains on the UK's Great Western Railway North Somerset Line, and was restored at Didcot to represent an original tower from the 1930s.

WHETHER IT WAS TO CONTROL a stop signal or to switch a passing train onto a different track, signal towers once served as the control hubs of a rail system. The towers ensured that trains operated safely over the correct route and in accordance with the schedule, and also provided the signalman with a warm and dry working environment. The earliest train signals were given by hand, or by the issuing of tokens, but over time signaling became more mechanical, using levers housed in the signal tower and positioned next to the track. The manually operated signal towers were often raised to accommodate the movement of the lower part of the levers, and to allow the signalman a clear view of the surrounding track. Today, with the advent of electric and electronic signaling technology, traditional signal towers have largely been replaced with centrally managed signaling control centers.

INTERIOR

The signal tower contains numerous levers set inside a frame mounted on a beam beneath the floor. The levers are painted according to their function. The large wheel operates the grade crossing, and the instruments on the shelf above the levers indicate whether or not different sections of the line are clear. The key token instrument offers a safety measure to ensure that no two trains are ever on the same line on a collision path.

1. Overview of signal tower interior 2. Token equipment for branch line (left) and main line (right) 3. Shelf containing block instruments and telegraph equipment 4. Close-up of three-position block instruments 5. Control levers: red for signals; blue for operating locks and gates; black for controling switches 6. Large wheel and levers for operating grade crossing 7. Wicket gate levers 8. Top of levers with release mechanisms 9. Brass plate denoting signal controlled by lever 10. Framed diagram of signaling system at Radstock North 11. Bell tapper to send coded messages to next signal box 12. Signal levers 13. Single line electric key token instrument 14. Hoops used to pass tokens to drivers 15. GWR clock 16. Lamp for signal tower operation at night 17. Wood-fired stove

EXTERIOR

The signal tower is built right next to the grade crossing and the train track, with the track connected to the tower by complex interlocking mechanics. Each lever inside the tower connects to a series of metal pulleys, chains, pivots, and rods that either change a signal, move a switch, or open and close a gate.

How Steam Locomotives Work

The power of steam has long been recognized as a potential energy source, and as early as the 1st century CE, steam-powered devices appeared in the writings of Hero of Alexandria. It wasn't until the dawn of the industrial era, however, that effective ways were found to harness steam power. In 1712 English ironmonger and inventor Thomas Newcomen developed a steam-powered pump to clear water from mines. Stationary engines such as Newcomen's became mobile when another English engineer, Richard Trevithick, began experimenting with high-pressure steam engines. These could be made small enough to be mounted on wheels and, for the first time, steam could be used for propulsion. Trevithick's first engine ran on roads, but in 1803 he built a steam locomotive for the Pen-y-Darren colliery that ran on iron track. Within 30 years, the rail revolution had begun, providing transportation for the masses, and steam was to power the world's railroads for more than a century.

CREATING STEAM POWER

To generate steam, hot gases pass from the firebox furnace along tubes that run through the boiler, where they are surrounded by water. The hot "fire tubes" boil the water, and steam collects at the top of the boiler. This is referred to as "saturated steam," and a regulator valve controls the rate at which it is fed into the main steam pipe. Superheater pipes then typically boost the steam temperature to give it even more energy before feeding it to the cylinders, where it expands to drive the pistons. Exhaust steam is released through the blast pipe to the chimney, helping to draw hot gases along the fire tubes.

KEY
- ■ Steam exhaust
- ■ Saturated steam
- ■ Superheated steam
- ■ Hot gases

Stoking the firebox
The fireman feeds coal into the firebox when the engine is running, but the fire will have been lit many hours before to raise the temperature slowly and avoid damaging the boiler.

STEAM LOCOMOTIVE COMPONENTS

The essential principles of steam power remained the same throughout the steam age, although locomotives grew more sophisticated. Early steam engines had just one fire tube, for example, but Stephenson's *Rocket* had 25 and later locomotives had 150 or more. Depending on its job—switching, hauling freight, or a passenger express—a locomotive had to deliver its power in different ways using more pistons or more drive wheels, but the basics remained largely the same.

Tender handbrake
Applies the tender brakes when the handle is turned

Coal space
Coal goes from here to the firehole via the fireman's shovel or an automatic feed system

TENDER

Water filler
Allows water tank to be filled from top of tender

Water tank
Supplies water to pipes that feed the boiler

Water float
Indicates water level in water tank

Brake rigging
Transmits pressure to the brake blocks on each wheel

STEAM PROPULSION

Water from the boiler is heated to produce steam, which is then superheated and transferred at high pressure via the steam pipe to the cylinder. Entering the cylinder through a valve, the high-pressure steam pushes a piston which, in turn, pushes the series of rods and pivots that turn the driving wheel, thus converting linear motion to rotation.

Steam feed High pressure steam in

Steam exhaust Low pressure steam out

Valve

Cylinder

Piston

1 Outward stroke High-pressure steam is fed via a valve into the front of the cylinder, where it expands and pushes the piston, which rotates the wheel by half a turn.

INSIDE THE CAB

Most steam locomotives had a crew of two: fireman and engineer. The engineer was in charge and controlled the locomotive using the regulator (which acts like a throttle), the reverser, and the brake. Watching his gauges and looking for trackside signals, the engineer regulated the train's speed. The fireman's duties were to maintain a good supply of steam by stoking the fire, and an adequate level of water by checking the gauge glass and using the injector control to force water from the tender into the boiler. With the engineer, he would also keep an eye out for trackside signals, especially on curves.

Steam chest pressure gauge

Vacuum brake pressure gauge

Blower control

Live steam water injector control

Vacuum brake lever

Regulator (throttle)

Reverser handle

Engineer's seat

Oil can warming tray

Boiler pressure gauge

Gauge glass shows boiler water level

Exhaust steam water injector control

Steam heating pressure gauge

Whistle lever

Cylinder drain-cock lever

Fireman's seat

Firebox door

Main steam pipe Carries saturated steam to the superheaters

Boiler The fire tubes are surrounded by water in the boiler which they heat to produce steam

Blast pipe Draws exhaust steam up the stack

Piston valve Supplies steam to either front or back of the cylinder

Fire tubes Carry hot gases from the firebox through the boiler to heat the water

Steam dome Directs rising steam into the main steam pipe

Regulator valve Controls the flow of saturated steam from the boiler

Stack Expels exhaust steam and boiler gases from the smokebox

Cylinder Collects steam from the steam pipe to push the piston

Firebox Supplies heat to the fire tubes

LOCOMOTIVE

Steam pipe Takes the superheated steam to the cylinder

Smokebox Collects hot gases that have passed through the boiler

Brake shoe Grips directly onto the wheel to slow the locomotive

Firebox grate The fireman shovels coal here from the cab

Drive wheel Linked to other drive wheels to receive power from the piston and give even traction to the rail

Superheater element Reheats saturated steam to produce superheated "dry" steam at high temperature

Coupling rod Links the drive wheels so that they all turn together

Connecting rod Transfers piston movement to the drive wheels

Piston Moves forward or backward inside cylinder when steam expands

Steam exhaust **Valve rod**

Piston rod

2 Exhaust The wheel is connected to the valve via a series of rods. These open the valve to allow the steam, which has now lost pressure, to escape.

Steam feed Steam enters back of cylinder

Piston Moves in return direction

3 Return stroke The movement of the valve also allows high-pressure steam to enter the back of the cylinder, allowing the return phase of the stroke to begin.

Steam exhaust

Piston Ready for next outward stroke

4 Exhaust Once the wheels have made another half turn, the valve allows spent steam to escape and fresh steam to enter, and the cycle begins again.

How Diesel Locomotives Work

The first diesel engine was demonstrated in 1893 by the German engineer Dr. Rudolf Diesel, who went on to build the first reliable example in 1897. A diesel engine works by drawing air into the cylinders and compressing it to increase its pressure and temperature. Diesel fuel is then injected into it, and the resulting combustion produces energy that pushes a piston, which drives a crankshaft. Different transmission systems (electric, mechanical, and hydraulic) are used to transfer the power from the crankshaft to the wheels. A diesel engine can be very powerful; those used in ships can be over 50,000 hp, though railroad applications are more typically 2,500–4,500 hp. Early diesel locomotives introduced in the 1930s and '40s were cheaper to operate than steam locomotives, especially where oil was plentiful, becasue they needed far less manpower. Today, diesel-powered trains are used worldwide, particularly on less busy lines where electrification is not economical.

DIESEL-ELECTRICS

Most diesel locomotives (and some diesel multiple units) have electric transmissions and are called "diesel-electric." In a diesel-electric, the power output in the diesel engine uses a transmission system to convert mechanical energy produced by the engine into electrical power. This is achieved by using the engine crankshaft to power a generator (more recently an alternator) to produce electricity. This electrical power in turn operates the traction motors, which are connected to the wheels or axles of the train. Diesel-electric locomotives are different from electric locomotives—they carry their own power plants rather than relying on an outside supply of electricity.

Diesel-electrics originally ran on DC (direct current) power supplied by a generator, but developments in technology in the 1960s allowed for the use of more reliable AC (alternating current) power supplied by an alternator instead of a DC generator. The AC power from the alternator was passed through a rectifier to transform it to DC electricity to power the traction motors. Advances in traction inverter technology in the 1980s and 1990s allowed the AC supply to power the motors directly, using a system known as three-phase supply.

Air intakes
Filtered intakes for air for the engine and other systems

Turbocharger
Generates extra horsepower from the engine using hot exhaust gases

Electric control cubicle
Contains electrical controls

Muffler
Reduces engine noise in the exhaust

Alternator
Converts rotary mechanical power from the engine to electricity to power the motors

Engine
Produces mechanical power through internal combustion

Rectifier
Converts the AC power output of the alternator to DC output

Radiator fan
Part of the cooling system to remove excess heat

Flexicoil suspension
Between locomotive body and trucks to minimize unsprung weight

Braking equipment cabinet
Contains electropneumatic braking equipment for whole train

Air compressor
Compresses air for use in braking and electrical cooling

Fuel tank
A diesel locomotive must carry enough diesel fuel to last each journey

Battery box
Contains batteries used to start the engine and operate onboard equipment

DIESEL-MECHANICALS

A mechanical transmission on a diesel locomotive consists of a direct mechanical link between the diesel engine and the wheels. There are two types of mechanism to achieve this. In a direct-drive mechanism, the engine is connected to the axles via driveshafts, differentials, and gearing. The second type is the coupling-rod drive, which is used on rigid locomotives that have no pivoting trucks. To maintain efficient adhesion, coupling rods are attached to the outer sides of the wheels of all the powered axles, powering all of the wheels at once.

Switcher
A switcher or shunter is a small railroad locomotive used for moving trains safely between storage yards and passenger stations. Shunters also assemble freight trains before a hauling locomotive takes over. Many shunters are diesel-mechanical locomotives as they do not need to be capable of high speed.

How it works
The diagram above shows how power is transferred from the diesel engine to the traction motors on the wheels, through the alternator and rectifier.

DIESEL-HYDRAULICS

Diesel-hydraulic locomotives have similarities to their diesel-mechanical cousins, but while most diesel-mechanical locomotives or diesel-mechanical multiple units are only capable of relatively slow speeds using low-powered engines, diesel-hydraulics are able to operate at higher speeds with much more powerful engines. This is because they have a torque converter instead of a gearbox. The torque converter contains a thick, viscous fluid inside a rotary impeller system to transfer power based on the amount of speed and power the engine is producing. German designers favored diesel-hydraulics after World War II, and large numbers were built; locomotives were even built for export as far afield as the US and Asia.

Torque converter
A torque converter contains hydraulic fluid that acts within rotating elements. One element, the impeller, is driven by the rotary power output of the diesel engine. The impeller blades drive the fluid onto the turbine blades, driving the turbine round and passing rotary force—or torque—on to the wheels. Extra torque is required when starting a locomotive; less torque is needed to maintain a constant speed.

Traction motor blower
Fan to cool down traction motors on this side of the locomotive

Luggage van
Space at rear of power car to store luggage

Truck
Specially designed for high-speed operation

Traction motor
Powers the train using electricity generated by the alternator; one attached to each axle

Main air reservoirs
Contain air for braking and other uses

KEY
- Driving impeller
- Turbine (driven)
- Fixed guide vanes
- → Flow of fluid

How Electric Locomotives Work

In Europe, electric trains were initially developed as a more efficient alternative to steam and early diesel locomotives. The first electric locomotive ran in 1879 in Berlin, Germany. However, much of the impetus for the switch to electric traction was driven by the increasing use of rail tunnels, especially in urban areas. In 1890, the first working underground system opened in London using electric locomotives, and electricity soon became the power supply of choice for subways, helped greatly by the introduction of multiple-unit train control in 1897. In the US, electrification of a main line was first used on a 4-mile (6.4-km) stretch of the Baltimore Belt Line of the Baltimore & Ohio Railroad, although electrification was confined to urban areas with dense traffic. The introduction of alternating current as a power supply allowed longer and heavier trains to be operated by electric locomotives and also increased their speed and efficiency.

ELECTRIC TRAINS

Like diesel-electric locomotives, electric trains employ electric motors to drive the wheels, but unlike diesel-electrics, electricity is generated externally at a power station. The current is picked up from catenaries (overhead cables) via a pantograph, or from a third rail. Since they do not carry their own power-generating equipment, electric locomotives have a better power-to-weight ratio and greater acceleration than their diesel-electric equivalents. This makes electric trains ideal for urban routes with multiple stops. They are also faster and quieter than diesel-powered trains. The world rail speed record is held by an electric train—a specially converted French TGV that achieved 357¼ mph (574.8 km/h) in 2007.

Electric locomotive components
For an electric locomotive that is powered via catenary, the pantograph picks up the power supply and transfers it to a transformer, where it is converted to the correct voltage to power the traction motors attached to each wheel. This power allows the locomotive to move.

Air-conditioning unit
Provides air conditioning for driver's cab and electrical equipment

Air reservoirs
Supplies air for traction motor blowers and other compressed-air-cooled electrical equipment

Motor blowers
Main rectifier
Compressor
Cooling fans
Circuit breaker
Catenary
Pantograph
3-phase AC motor
Auxiliary rectifier
Main transformer
Main inverter
3-phase AC motor
Auxiliary inverter

How it works
In the three-phase AC electric locomotive above, the AC power supply is converted to lower-voltage DC power by the transformer and rectifiers. Inverters then convert this power back to AC—but at the same lower voltage—to supply power to the motors.

KEY
■ High-voltage AC current from catenary
■ Converted lower-voltage DC current
■ Converted lower-voltage AC current

Smoothing choke
Smooths the DC electric supply to ensure consistency of supply to the motors

Air compressors
Feed the traction motor blowers that help keep the engine cool

THIRD RAIL

Many subway and light rail systems use a third power rail as a method of power supply because it is cheaper to install than overhead lines and is relatively efficient. A shoe extending from the train makes contact with the power rail and conducts electricity to the train. The system has the advantage that many trains can use it at the same time, disengaging when they no longer need power. The power rail carries a high current that is potentially fatal to humans and animals that come in contact with it, so measures are taken to minimize the risk of contact, especially in stations and depots.

Rail
Power rail
Insulator
Protective cover
Tie

Third-rail layout
The power rail lies on insulators mounted on ties, and sits alongside the running rails used by the wheels of the train.

OVERHEAD LINES

Electric trains that collect their current from catenaries (overhead cables) use a power-collector device such as a pantograph, bow collector, or trolley pole. The power collector is in contact with the lowest overhead wire—the contact wire. Normally made from copper or aluminum, the contact wire is designed to carry several thousand amps of current while remaining in line with the track and withstanding hostile weather conditions. The mechanics of power-supply wiring is not as simple as it looks. The contact wire's tension has to be kept constant; to negotiate curves in the route, for example, the wire has to be held in tension horizontally while it is pulled laterally. The overhead wire is deliberately mounted in a zigzag pattern to avoid wearing holes in the pantograph.

Insulator

Booster transformer

Contact wire

Mast

Dropper

Catenary

Running rail

Pantograph
The pantograph is kept in contact with the overhead line using a spring or an air pressure device. Its contact strips are designed so that they do not get hooked up over the top of the contact wire as the train moves along.

Main control cabinet
Contains thyristor-based controls to convert AC supply to DC required by traction motors

Rheostatic brake unit
Contains brake grid resistors that dissipate heat generated by the traction motors

Pantograph
Collects power from overhead cables using carbon-tipped head

Insulator
Protects locomotive and crew from high-voltage power collected from the pantograph

Motor contactor cubicle
Controls and regulates power fed to the traction motors

Field control cubicle
Contains electrical control equipment for this end of the locomotive

Main air reservoir
Air supply for train braking system

Transformer (behind battery box)
Reduces voltage of electricity supplied by overhead cables to a suitable voltage for traction motors

Traction motor
Provides the propulsion to move the locomotive; attached to each axle with integrated gearbox

Traction motor blower
Provides air cooling for the motors mounted on the truck below

Three-axle truck
All axles have traction motors attached and are powered

Shoe contact
Trains are equipped with a "shoe" that collects current from the power rail. The simplest design is known as the "top contact," with the pick-up shoe sliding along the top part of the power rail. However, the smallest amount of snow or ice on the exposed rail can render it ineffective. Side contact offers more protection from the elements, but bottom contact is superior because it makes contact with most of the rail and is unaffected by bad weather.

Shoe on train

Power rail

Insulator

Sleeper

TOP CONTACT

Protective cover

Shoe on train

Power rail

Insulator

Tie

TOP CONTACT WITH COVER

Fastener

Protective cover

Shoe on train

Power rail

Tie

SIDE CONTACT

Protective cover

Power rail

Fastener

Shoe on train

Tie

BOTTOM CONTACT

Glossary

Adhesion
The frictional grip between the wheels of a locomotive and the rail of a track, which is affected by axle weight. Particularly important when a locomotive is starting.

Air brake
A braking system that uses compressed air as its operating medium. To apply the brake, the compressed air is released into a cylinder, pushing a piston and spring that push the brake block against the wheel.

Air cushion
A "spring" of air used in modern suspension systems.

Alternating current (AC)
An electric current that reverses its direction of flow rapidly at regular intervals. The rate at which it reverses per second is the frequency, and is calculated in cycles, or hertz (Hz). *See also* **Direct current (DC)**

Alternator
An electromechanical device that converts mechanical energy to electrical energy in the form of alternating current (AC). Used in diesel-electric and electric locomotives.

Articulated locomotive
A locomotive (often steam) with two or more engine units mounted on the same frame but pivoted so that they can move independently of each other. This allows them to transition through curves despite a long wheelbase.

Articulated train
An interconnected train set with pairs of cars linked by single, pivoting trucks.

Ashpan
Located beneath the firebox of a coal-powered steam locomotive, this pan collects the ash and cinders that fall through the grate of the firebox.

Atlantic
A steam locomotive with a wheel arrangement of 4-4-2—four leading wheels on two axles, four powered and coupled wheels, and two trailing wheels. First seen in 1880, it was also called a Milwaukee, after the Milwaukee Road, which used the type for its high-speed passenger operations.

Axlebox
A metal casing housing the bearing in which the end of an axle rotates.

Axle load
The fraction of a vehicle's weight that is carried by a given axle. Tracks are designed to carry a maximum axle load.

Ballast
The bed of stone, gravel, or cinders on which a rail track is laid. Ties are bedded into the ballast to support the rails. *See also* **Blanket, Formation, Subgrade**

Banker *see* **Helper**

Bar frame locomotive
A lightweight steam locomotive originally developed by Edward Bury in 1838, which had a frame made of bars rather than plates. The bar-frame type was adopted as standard in the US.

Bell code
A language using bell signals to describe trains used by signalers to receive and pass on trains.

Bell tapper
A device used to tap out bell signals between signalers.

Big end
The crank pin end of a connecting rod, larger than the crosshead end.

Blanket
An optional layer in the formation of track, the blanket is made of coarse material, and supports the layer of ballast. *See also* **Ballast, Formation, Subgrade**

Blastpipe
A pipe that conveys exhaust steam from the cylinders up the stack of a steam locomotive. This creates a partial vacuum, increasing the flow of air passing through the firebox.

Block
In signaling terms, a section of track that sits between two signals. Trains cannot enter the block if the first signal is "stop."

Bo-Bo
A common axle configuration that describes a locomotive that has two groups of twin-set powered axles. *See also* **Co-Co, Wheel arrangement**

Bogie *see* **Truck**

Boiler
The part of a steam engine in which steam is produced and circulates. The boiler must be filled with water almost to the top. The water is generally heated by fire tubes, producing steam, which builds to a high pressure. The fireman ensures the boiler is sufficiently filled with water.

Boilerman *see* **Fireman**

Boxcar
A flat-bottomed freight car with sliding doors on each side.

Brake
A locomotive has a set of brakes to slow itself down. It is normally equipped with an additional control that engages brakes along the length of the train via the brake rigging. Brakes are activated by either air, steam, or a vacuum. *See also* **Air brake, Vacuum brake**

Brake block
The friction material that is pressed against a wheel to slow a train down when the brake is applied.

Brake rigging
The system of rods and levers that connect the brake controls to the brake blocks on each wheel.

Brake van *see* **Caboose**

Branch line
A secondary rail line that branches off a main line, serving local stations.

Broad gauge
Any gauge in which rails are spaced more widely than the standard gauge of 4 ft 8½ in (1,435 mm); for example, Isambard Kingdom Brunel's 7-ft-¼-in (2.14-m) gauge.

Bullhead rail
A type of rail in which the top half of the rail mirrors the bottom half. This design was intended to make rails last longer. Once the running side is worn out, the rail can be turned over and reused.

Bumper
A device that cushions the impact of rail vehicles against each other.

Bumper post
The structure at the end of a track that halts a train from traveling any farther.

Bunker
When a steam locomotive was not followed by a tender, this enclosure at the back of the cab was used to store coal.

Bus connector
On an electric multiple unit train, the equipment that transfers electricity supplied by the catenary from one unit to the next.

Cab
The control room of a locomotive, housing the engine crew.

Cabin car
A rail car used by railroad workers to monitor track conditions. It is usually attached to the end of a train.

Caboose
A railcar that provides braking power for freight trains and accommodation for the conductor. Also known as a brake van.

Cant *see* **Superelevation**

Car, carriage, coach
Various terms that describe a passenger-carrying rail vehicle.

Catenary
Originally referring to just the wire that supported the conductor wire of an overhead electrification system, the term catenary is now used to apply to the entire overhead wire arrangement. Also known as overhead lines and overhead wires.

Chimney *see* **Stack**

Class
A group of locomotives built to a common design. Can also refer to the level of passenger comfort and service provided on a particular train or car, e.g., first class.

Class 1 railroad
A US mainline railroad that has annual carrier operating revenues of more than $250 million.

Classification light *see* **Marker light**

Classification yard
A place where freight trains are assembled, or where freight cars for different destinations are moved to the correct train. Also known as a marshaling yard.

Co-Co
Refers to any diesel or electric locomotive that has two triple-sets of powered axles. *See also* **Bo-Bo, Wheel arrangement**

Coal space
The portion of a tender that carries coal to fuel the firebox. The rest of the tender carries water for the boiler.

Collector shoe
A power collection device attached to an electric train that picks up electricity from an electrified third rail that runs alongside the running track.

Compound locomotive
A steam locomotive that uses two sets of cylinders, the second powered by exhaust steam from the first.

Compression ignition
The process of using heat from compression to ignite and burn fuel in an internal combustion engine. Compression ignition engines are known as diesel engines, and are contrasted with spark ignition engines that use a spark plug to ignite fuel. *See also* **Diesel**

Conductor
A member of a train's crew who performs ticketing duties. The conductor looks after packages and other freight, and may also be responsible for the brakes. Known in the UK as a guard, though the term "conductor" is increasingly used there.

Conjugated valve gear
The operation of a valve on a steam locomotive cylinder by means of levers driven by the motion of the valve gear on two other cylinders. Used by Sir Herbert Nigel Gresley on the 3-cylinder locomotives he designed for the Great Northern and the LNER in the UK.

Connecting rod
On a steam engine, a connecting rod links the piston rods to the crankpins of the drive wheels. In some early electric locomotives, the connecting rods linked the crankshaft with the drive wheels.

Consolidation
A locomotive with a 2-8-0 wheel arrangement. It has two leading wheels on one axle, followed by eight powered and coupled drive wheels on four axles. Introduced in the 1860s, it was popular in the US and Europe as a freight hauler.

Container
A standardized metal freight box that can be packed with goods, sealed, and then transported by specially adapted trains, trucks, and ships.

Coupler, Coupling
The mechanism for connecting rail vehicles together. Methods are standardized across a single railroad to allow any rolling stock to be coupled together. Known in the US as a coupler, and in the UK as a coupling.

Coupling rods
The drive wheels along both sides of a steam locomotive are linked together by coupling rods, also known as side rods. Coupling the wheels spreads the power and reduces the possibility that the wheels will slip.

Cowcatcher
A sloping plate or grid on the front of a locomotive, designed to push obstructions off the track. Also known as a pilot.

Crank
The part of a steam locomotive that transmits power from the piston to the drive wheels.

Crankpin
A large steel pin that is pressed into the wheel center. On steam engines, the drive wheels are driven by rods that transmit force to the wheels through the crankpins.

Crankshaft
In steam locomotives, a shaft that acts upon cranks in order to convert the linear motion of the piston into rotary motion. This rotary motion drives the wheels.

Crosshead
The point of connection between the piston and the connecting rod that, along with the slidebars, keeps the piston rod in line as it moves in and out of the cylinder.

Cutting
A channel dug through a hillside to enable a railroad track to maintain a shallow grade.

Cylinder
An enclosed chamber in which a piston moves to produce power that is transmitted to the wheels. On a steam locomotive, the piston is moved by the force of high-pressure steam acting against it.

Diesel
Unlike gasoline engines, diesel engines use compressed air, rather than a spark, to ignite the oil that fuels them. On a locomotive, the transmission of power from a diesel engine to the wheels may be by electric, mechanical, or hydraulic means. *See also* **Compression ignition**

Diesel-electric
Any locomotive, multiple unit, or railcar that utilizes the diesel-electric system. In a diesel-electric, mechanical power generated by combustion is then converted into an electric charge in a generator or alternator, and this electricity is used to power motors that drive the axles.

Diesel-hydraulic
Any locomotive, multiple unit, or railcar that utilizes the diesel-hydraulic system. In a diesel-mechanical, mechanical power generated by combustion is passed through a torque converter that transfers power to the wheels based on the amount of speed and power the engine is producing.

Diesel-mechanical
Any locomotive, multiple unit, or railcar that utilizes the diesel-mechanical system. In a diesel-mechanical, mechanical power generated by combustion is transferred directly to the wheels by means of driveshafts, gearing, and differentials.

Direct current (DC)
An electric current that flows in a constant direction. Alternating current (AC) has significant advantages over direct current in terms of transforming and transmission.

Double-heading
The use of two locomotives, with separate crews, at the head of a train.

Drive wheels
The powered or driven wheels of a locomotive that provide traction.

Dynamic braking
In electric and diesel-electric locomotives and multiple units, the electric traction motors can be used as generators that act as brakes to slow down the train. Excess energy may be dissipated as heat through brake grid resistors (this is known as rheostatic braking). On an electric train, the excess energy may also be absorbed back into the power supply system (this is known as regenerative braking).

Dynamometer
A device (also called a dyno) used for measuring force, torque, or power. On railroads, dynamometer cars are used to measure a locomotive's speed.

Ejector
Part of a vacuum brake system. The ejector evacuates the brake pipe to create a vacuum, which releases the brakes.

Electrics
Refers to all locomotives, multiple-unit trains, and railcars that draw electric power for traction from an external source. The electric supply is drawn from a conductor rail beside the track or from catenaries.

Elevated railway
A railway built on raised platforms. Examples are the former Liverpool Overhead Railway in the UK and part of the New York Subway in the US.

Embankment
A raised pathway across a depression in the landscape that enables a rail track to maintain a shallow gradient.

Engine
The power source of a locomotive, driven by steam, electricity, or diesel. Steam locomotives may also be referred to as steam engines.

Exhaust
Used steam and combusted gases produced by either a steam or a diesel locomotive.

Express train
A train that stops only at certain stations on its route in order to reach its final destination faster.

Firebox
The section at the rear of a steam locomotive boiler that houses the fire that heats the water in the boiler. Fuel is fed into the firebox from the cab, and the generated heat is fed through the boiler by the fire tubes.

Firehole
The aperture in the firebox of a steam locomotive through which coal or other fuel is fed by the fireman.

Fireman
A worker responsible for keeping the firebox of an engine fed with coal or other fuel. Also known as a stoker or boilerman.

Fire tubes
Tubes running between a steam locomotive's firebox and smokebox. Hot gases drawn through the fire tubes heat the water surrounding the tubes.

Flange
The projecting lip on the inside edge of a wheel that guides the wheel along a rail.

Flat-bottomed rail
Standard rail used today, which takes the form of a T with a wide, flat base.

Footplate
The floor of a locomotive engineer's cab where the crew stands. Footplate can also refer to the entire cab.

Formation
The substructure of a track on which the ties and rails are laid. *See also* **Ballast**, **Blanket**, **Subgrade**

Freight
A term used to describe trains transporting finished goods and raw materials. It can also refer to the load of materials or products that are being carried.

Funicular railway
Used on tram, cliff, and industrial lines, funicular railways use cables or chains to move vehicles up and down slopes.

Gangway
A flexible structure provided at the ends of coaches to provide access from one coach to another.

Garratt locomotive
An articulated steam locomotive with a boiler in a central frame and two engines on separate frames at each end.

Gas turbine
A type of internal combustion engine that uses high-temperature, high-pressure gas to generate energy. Both US and Russian railroads are now experimenting with gas turbine-electric locomotives (GTELs), which use a gas turbine to drive an electric generator or alternator.

Gauge
The distance between the inside running edges of the rails of a track. Many gauges are used in different countries and on different railroads. Also denotes a visual display of readings for steam, pressure, etc.

Gauge glass
A vertical glass tube in a steam locomotive cab that indicates the water level in the boiler and firebox.

Generator
An electromechanical device that converts mechanical energy to electrical energy in the form of direct current (DC).

Gondola
An open-top piece of rolling stock used to transport loose materials such as ore and coal. Also called an open wagon.

Grade, Gradient
The slope of a track. Known in the US as grade and in the UK as gradient.

Grade crossing
A location where a railroad crosses a road at the same elevation. Also known as a grade crossing or simply a railroad crossing.

Grate
A grille of firebars at the base of a firebox upon which the fire rests. The gaps in the grille allow in air to assist the fire.

Guard *see* **Conductor**

Handcar
A small, open rail vehicle propelled by its passengers, often by means of a hand pump. Also known as a pump trolley.

Helper
An extra locomotive that is coupled to a train to help it climb a steep section of track. Also known as a banker.

Horsepower (hp)
A unit of power equal to 745.7 watts or 550 foot-pounds per second. Used to express the power produced by steam, diesel, or electric locomotives.

Hot box
Term for an axlebox that has overheated due to inadequate lubrication or too heavy a load.

Injector
A device that feeds water into the boiler of a steam locomotive against the pressure of steam in that boiler.

Interlocking tower *see* **Signal box**

Intermodal
A term used to describe a freight container that can be transferred from one mode of transport to another, such as from a train to a lorry or a ship.

Inverter
A piece of electrical equipment on a diesel-electric or electric locomotive that converts direct current (DC) power supply into an alternating current (AC) supply.

Jacobs bogie
Designed by German railway engineer Wilhelm Jakobs, this is a type of truck used on articulated railcars and tram vehicles. The truck is placed between two car body sections, rather than underneath, so that the weight of each car is spread on one half of the truck.

Journal box
The housing in which the end of an axle turns on a bearing.

Kriegslok
Short for *Kriegslokomotive*, this is a German war locomotive. Built in large numbers during World War II, they were cheap and easy to build, easy to maintain, and could withstand extreme weather conditions.

Leading wheel
A wheel located in front of the drive wheels of a steam locomotive that provides support but is itself unpowered.

Level crossing *see* **Grade crossing**

Level junction
A railroad junction where multiple lines intersect, crossing the path of oncoming rail traffic at the same elevation.

Light rail
A form of rail typically operating within urban environments. Light rail vehicles (LRVs) include streetcars and trams.

Link valve gear
A design of valve gear, designed at the Stephensons' locomotive works in 1842.

Livery
Distinctive colors, insignia, and other cosmetic design features of a rail vehicle.

Loading gauge
The dimensions that a rail vehicle must not exceed, to avoid collisions with trackside objects and structures. Different countries have different loading gauges.

Locomotive
A wheeled vehicle used for pulling trains. Steam and diesel locomotives generate their own power, while electric locomotives collect electricity from an external source.

Maglev train
A train that works by being levitated above and propelled over special tracks by electromagnetic force. Maglevs produce virtually no friction, and are very quiet in operation at high speed.

Main line
An important rail line, often running between major towns or cities.

Marker light
Particularly in the US, a light that was used to signal the status of the train to other drivers. Green marker lights indicated a regularly scheduled train, white marker lights indicated an extra train, and red marker lights attached to the final car indicated the end of the train. Red lights are still used in tail lights around the world today. Also known as a classification light.

Marshaling yard see **Classification yard**

Meter gauge
A railroad track with the inside of its rails 3 ft 3 in (1 m) apart.

Metro
Internationally, a name that is popularly used for an underground rapid transit system—a type of high-capacity rail mass transit in urban areas. Generally known as a subway in the US. Each system has its own name, such as London Underground, New York Subway, and Paris Metro.

Monorail
A railway system based on a single rail. A monorail is often elevated above the ground, and built in urban areas.

Motion
In railway terminology, the collective term for the piston rods, connecting rods, and valve gear of a locomotive.

Motive power depot see **Running shed**

Multitube boiler
A locomotive boiler with multiple tubes, which revolutionized steam locomotive design. Stephenson's *Rocket* was the first engine to have a multitube boiler—with 25 copper tubes instead of a single or twin flue.

Multiple unit (MU)
A term used in diesel and electric traction that refers to the semipermanent coupling of several powered and unpowered vehicles to form a single train.

Narrow gauge
Any railroad with a gauge narrower than the standard 4 ft 8½ in (1,435 mm).

Oil firing
A method of firing a steam locomotive using oil as fuel.

Open wagon see **Gondola**

Overhead lines or **Overhead wires** see **Catenary**

Pacific
A locomotive with a wheel arrangement of 4-6-2. It has four leading wheels on two axles, six powered and coupled drive wheels on three axles, and two trailing wheels on one axle. The Pacific was a common type of steam passenger locomotive during the first half of the 20th century.

Pantograph
An assembly on the roof of an electric locomotive or electric multiple-unit power car that draws current from an overhead wire (catenary). Also known as a current collector.

Passenger train
A train with cars intended to transport people rather than freight. These trains travel between stations at which passengers may embark or disembark.

Passing loop, passing siding
A position on a single-track railroad where trains traveling in opposite directions can pass each other. Known as a passing loop in the UK and as a passing siding in the US.

Permanent way
The rails, ties, and subgrade of a rail line. The term comes from the fact that temporary lines were laid during railroad construction, and then replaced by a "permanent way."

Pilot see **Cowcatcher**

Piston
The cylindrical assembly that moves back and forth inside a cylinder of a steam or diesel engine. The movement of the piston provides mechanical power, which is transferred by various means to the wheels.

Piston rod
The rod linking the piston in a cylinder with the crosshead.

Points see **Switch**

Pullman car
A luxury rail car. Pullmans were initially introduced in the US by George Pullman in 1865 as sleeping cars on long-distance trains.

Pump trolley see **Handcar**

Rack railway
A railway with an additional toothed rack-rail. A train or locomotive running on the railway is equipped with a cog that lines with the teeth on the rail, enabling it to climb slopes that would be impossible for a normal train.

Railcar, railmotor
A self-propelled passenger vehicle, usually with the engine located under the floor.

Railroad time
Before the introduction of railroad schedules, different places in the same country often had their own local time.

In the 1840s, railroads began to introduce a standardized railroad time to avoid confusion caused by local time differences.

Rectifier
A piece of electrical equipment on a diesel-electric or electric locomotive that converts an AC power supply into a DC power supply. They are also used alongside tracks to convert traction current.

Regenerative brake see **Dynamic braking**

Regulator see **Throttle**

Reverser
Mechanism with a wheel or lever that controls the forward and reverse motion of a steam locomotive.

Rheostatic brake see **Dynamic braking**

Rolling stock
The collection of vehicles that a rail company runs on its railroad.

ROD
Stands for the Railway Operating Division of the British Royal Engineers, who maintained the railroads in theaters of war during World War I.

Running board, running plate
The footway around a locomotive's engine compartment or boiler.

Running gear
The parts involved in the movement of an engine. Includes wheels, axles, axleboxes, bearings, and springs.

Running shed
An old name for a motive power depot, where locomotives are stored, repaired, and maintained when not in use.

Saddle tank
A tank locomotive that has the water tank mounted on top of the boiler.

Safety valves
In a steam locomotive boiler, relief valves that are set to lift automatically to allow steam to escape if the boiler pressure exceeds a set limit.

Saloon
A luxurious rail car used as a lounge, or with private accommodations.

Sandbox see **Sanding**

Sanding
The application of sand between the wheels and the rails to increase grip and prevent wheelslip. The sand is piped from a sandbox, which is often situated on top of the boiler.

Saturated steam
Steam that has yet to be superheated to remove any remaining water droplets. Also known as "wet steam."

Semaphore signaling
A system that relies on pivoting arms to relay a signal to drivers. The angle of each pivoting arm tells the driver whether the signal is "stop," "caution," or "all clear."

Shoe see **Collector shoe**

Shunter see **Switcher**

Shuttle
A rail service that operates between two stations, often without intermediate stops. A common use of shuttle services is to take passengers between airport terminals, or from an airport to a downtown location.

Side rods see **Coupling rods**

Siding
A section of track off the main line used for storing rolling stock.

Signal
A mechanical or electronic fixed unit with an arm or a light that indicates whether a train should stop, go, or use caution.

Signaling token
A token used in old signaling systems. The token was collected by the train's crew at the beginning of a block of track. The token was returned at the other end of the block. This system ensured that at any time, only one train would be allowed within the block of track.

Signaler
A signal maintenance worker.

Signal box see **Tower**

Sleeper
A coach or train with beds for passengers on overnight or long-distance journeys. Also another name for a railroad tie.

Sleeping car
A car with beds where passengers can sleep while traveling. Sleeping cars were first introduced in the US in the 1830s.

Slidebars
On a steam locomotive, slidebars combine with the crosshead to guide the movement of piston rods.

Slip coach
A coach that could be uncoupled from a moving express train and braked to a halt at a station. This allowed passengers to disembark without stopping the main train.

Smoke deflectors
Metal sheets attached to the smokebox to funnel air upward, forcing smoke and steam emitted from the stack away from the cab to improve visibility.

Smokebox
The leading section of a steam locomotive boiler assembly that houses the main steam pipes to the cylinders, the blastpipe, the stack, and the ends of the firetubes. Ash drawn through the firetubes collects here.

Spiral
A railroad formation where tracks cross over themselves as they ascend a mountain.

Splasher
A semicircular guard used to enclose the top section of a large-diameter drive wheel. Often installed when a wheel protrudes above the running board of a locomotive.

Stack
The opening in the top of the smokebox through which exhaust gases and steam escape. Also called a smokestack or chimney.

Standard gauge
Rails spaced 1,435 mm (4 ft 8½ in) apart. Standard gauge is the most commonly used gauge worldwide. Designed by Robert Stephenson for the first intercity railroads, it is also known as Stephenson's gauge.

Steam chest
The internal part of a locomotive's cylinder block where the valve chamber connects with the steam supply and exhaust pipes.

Steam dome
A chamber on top of the barrel of a steam locomotive's boiler where superheated steam collects and is directed to the cylinders through the steam pipe.

Steam locomotion
Steam locomotion is founded on the principle that when water is heated above its boiling point, it turns to steam and its volume becomes 1,700 times greater. If this expansion takes place within a sealed vessel such as a boiler, the pressure of the steam will become a source of energy.

Steam pipe
The pipe that connects the steam dome to the steam chest in the cylinder block.

Stoker see **Fireman**

Streamliner
A locomotive or train set that incorporates streamlining into its shape to provide reduced air resistance.

Subgrade
Ground prepared to give a consistent gradient to tracks that will be laid above it. See also **Ballast**, **Blanket**, **Formation**

Subway see **Metro**

Supercharging
A method of introducing more air into the cylinders of a diesel engine, through the use of a turbocharger to force air through the inlet valves at higher than atmospheric pressure.

Superelevation
The angle of elevation of a rail, relative to vertical or to its partner rail. Also known as cant.

Superheated steam
Steam that has been raised in temperature and volume by adding extra heat as it passes between the boiler and the cylinders. This dries the steam by turning the remaining water droplets into gas, thus delivering more power.

Switch
A trackwork mechanism at the point where two tracks diverge that allows a train to move from one track to another.

Switcher
A small locomotive used for moving trucks or wagons around in a marshaling yard. Also known as a shunter.

Tail light
The lamp at the rear of a train, In the UK, a train is not complete without a red rear warning light. See also **Marker light**

Tank locomotive
A steam engine that carries its fuel and water on its own chassis rather than in a

separate tender. The water is often held in side tanks or in saddletanks that encase the boiler.

Telegraph (electric)
A communication system developed in the 1830s that used electrical impulses traveling through wires to send messages. It became the standard instrument of railroad communication worldwide.

Tender
A vehicle, attached to a steam engine, that carries the fuel and water.

Third rail
A system that provides an electric train with power through a conducting third rail set alongside the running tracks. The power is collected via a shoe attached to the train.

Three-phase system
A system that enables a steady supply of AC current without fluctuations to power traction motors, enabling higher traction power to be achieved.

Throttle
A lever used by the engineer of a steam locomotive to control the supply of steam to the cylinders. Also known as a regulator.

Tie
The cross-piece supporting the rails, made out of wood, concrete, or steel. Early railroads also used stone blocks as ties.

Tilting train
A train that can lean into bends, enabling it to travel faster around curves without passenger discomfort.

Tower
A control room in which the movement of trains is controlled by means of signals and blocks, ensuring that trains travel safely and on schedule. Also known as a signal box.

Towerman
A person employed by a railroad to manage and operate the switches and signals on a section of track from a signal box.

Track
The permanent fixtures of rails, ballast, fastenings, and underlying substrate that provide a runway for the wheels of a train.

Traction
In railroad terms, a force that relies on friction between a wheel and a rail to generate motion. See also **Adhesion**

Traction motor
An electric motor that uses incoming electrical energy to power the axles. Used in both diesel-electric and electric traction.

Tractive effort
A measure of a locomotive's pulling power; the effort that it can exert in moving a train from standstill. This force is calculated by measuring the energy the locomotive exerts on the rails. See also **Traction**

Trailer, trailer car
A passenger vehicle in a multiple unit that has no power traction equipment, and is powered by the vehicles that are attached to it.

Trailing wheel
A wheel located behind the drive wheels of a steam locomotive that provides support but is unpowered.

Train
Passenger or freight vehicles coupled together and traveling as one unit along a rail line. Trains can be self-propelled or locomotive-hauled.

Transfer
A train station where passengers can transfer from one train to another that follows a different route.

Transmission
In a diesel locomotive, the method by which power is transmitted from an engine to an axle or the wheels. Transmission may be electrical, hydraulic, or mechanical.

Truck
A set of pivoted wheels attached to suspension components placed at the front or rear of a locomotive to give guidance and added support. Sometimes called a bogie. The term can also mean a small rail wagon.

Turntable
A device for rotating rail vehicles so they can travel back in the direction they came from. Largely obsolete today.

Twin-track railroad
A railroad that runs two tracks along the same line, each track taking trains in opposite directions, rather than both directions being serviced by a single track.

Underground see **Metro**

USATC
An abbreviation that stands for United States Army Transportation Corps. Locomotives built in the US for the USATC were shipped to Europe for use by the Allies in World War II.

Vacuum brake
A type of brake that is held off by a partial vacuum and applied when air is let into the system. Vacuum brakes were used in the UK because, unlike air brakes, they did not require a separate pump.

Valve
In a steam locomotive, valves coordinate the movement of steam into and out of the cylinders. In a diesel engine, valves control fuel intake and expulsion of exhaust gases.

Valve gear
Linkages that connect the valves of a steam locomotive and control the movement of the valves.

Vertical cylinder
Vertically mounted cylinders used in early locomotives such as the Stephensons' *Locomotion No.1* and, later, in specialized forms of shunting engines and narrow-gauge locomotives.

Wagon
A general term for a rail vehicle that carries freight.

Walschaerts valve gear
A form of link motion valve gear first patented in 1844 by Egide Walschaerts, a Belgian engineer. It became very

widespread in Europe, being easier to maintain and lighter than Stephenson's link valve gear. It first appeared in the US in 1876 and was also widely adopted there.

Water column, water plug
A hollow pole equipped with a hose and connected to a water supply for filling locomotive water tanks. Water columns may be installed on cranes with movable arms to allow water to be supplied to locomotives on either of two adjacent tracks. Known in the UK as a water column, and in the US as a water plug.

Westinghouse brake
A widely used automatic air brake invented in the 1870s by US engineer George Westinghouse. Universally adopted in the US, it was also developed worldwide.

Wet steam see **Saturated steam**

Wheel arrangement
A method of classifying locomotives by the distribution of different types of wheels. For steam locomotives, Whyte notation is a common system. Diesel and electric locomotives and powered cars are categorized by the number of powered and unpowered axles that they have. The unpowered axles, which often carry the leading and the trailing wheels, are listed numerically, while the powered axles supporting the drive wheels are given an alphabetical description. See also **Bo-Bo**, **Co-Co**, and **Whyte notation**.

Wheel unit see **Truck**

Wheelset
An assembly that consists of two wheels attached to an axle on a rail vehicle.

Whyte notation
A classification of steam locomotive wheel arrangements that is based on the number of leading, drive, and trailing wheels. For example, a wheel arrangement of 4-4-0 would denote a locomotive with four leading wheels, four drive wheels, and no trailing wheels.

Yard
An area off the main line used for storing, sorting, loading, and unloading vehicles. Many rail yards are located at strategic points along a main line. Large yards may have a tower to control operations.

Index

All general page references are given in *italics*.
References in **bold** refer to main entries.

Acknowledgments

Dorling Kindersley would like to thank Tony Streeter for all his time, assistance, and support throughout the making of this book.

General Consultant Tony Streeter is a journalist and editor who writes across the rail spectrum from steam to modern railways and international light rail. A former long-term editor of UK's *Steam Railway Magazine*, he has traveled by, and written about, rail in Russia, China, India, Canada, and Eastern and Western Europe.

Tony Streeter would like to thank the many people who assisted in the making of this book, including: Pip Dunn, Peter Johnson, Anthony Coulls, Tim Bryan, Bernd Seiler, Richard Croucher, Paul Chancellor, Brian Stephenson, Marek Ciesielski, Robin Garn, Jacques Daffis, Uwe Hüttner, and Peter Weißhahn.

The publisher would like to thank the following people for their help with making the book: Steve Crozier at Butterfly Creative Solutions for color retouching; Simon Mumford for cartography; Phil Gamble for illustrations; Sonia Charbonnier for technical support; Nicola Hodgson for additional text contributions; Tejaswita Payal, Suparna Sengupta, Sreshtha Bhattacharya, and Neha Pande at DK Delhi for editorial assistance; Neha Sharma, Shruti Singhal, and Upasana Sharma at DK Delhi for design assistance; Joanna Chisholm for proofreading; Helen Peters for the index. The publishers would also like to extend a special thanks to contributors Keith Fender and Julian Holland, whose assistance throughout the project was invaluable.

The Publisher would also like to thank the following museums, companies, and individuals for their generosity in allowing Dorling Kindersley access to their rail vehicles and equipment for photography:

8 201, Dampf-Plus GmbH
Moosglöckchenweg 10, 80995 München, Germany
www.zugparty.de
With special thanks to Christian Goldschagg

Adrian Shooter
(Owner of DHR B Class No.19 and its carriages)

Ashford Depot
Station Road,
Ashford, TN23 1EZ, UK
With special thanks to Nigel King and Mark Fitzgerald

B&O Railroad Museum
901, West Pratt Street,
Baltimore,
MD 21223, US
www.borail.org
With special thanks to David Shackelford, Ryan McPherson, and Jane Harper

Didcot Railway Centre
Didcot Parkway Station,
Didcot, Oxfordshire,
OX11 7NJ, UK
www.didcotrailwaycentre.org.uk
With special thanks to Roger Orchard, Peter Rance, and Frank Dumbleton

Eisenbahnfreunde Traditionsbahnbettriebswerk Strasssfurt e.V.
Guestener Weg,
39418 Strassfurt, Germany
www.efsft.de
With special thanks to Uwe Hüttner

Ffestiniog & Welsh Highland Railways
Porthmadog,
LL49 9NF, UK
www.festrail.co.uk
With special thanks to Andrew Thomas and Chris Parry

Hitachi Rail Europe Limited
40 Holborn Viaduct,
London, EC1N 2PB, UK
www.hitachirail-eu.com
With special thanks to Daniela Karthaus

HSB, Harzer Schmalspurbahner
Friedrichstrasse 151,
38855, Wernigerode, Germany
www.hsb-wr.de
With special thanks to Bernd Seiler

National Railway Museum (NRM York)
Leeman Road,
York, YO26 4XJ, UK
www.nrm.org.uk
With special thanks to Chris Hanley

National Railway Museum
Chanakyapuri,
New Delhi, 110021, India
With special thanks to Uday Singh Mina, Director

Northern Railway
D.R.M. Office, State Entry Road, New Delhi - 110055, India
www.nr.indianrailways.gov.in
With special thanks to Rajesh Kumar, Sr. DME/Power/Delhi

Palace on Wheels
Rajasthan Tourism Development Corporation Ltd.
Ground floor, Bikaner House,
Pandara Road, New Delhi - 110011, India
www.rtdc.in
www.thepalaceonwheels.com
With special thanks to Pramod Sharma, General Manager, Rajasthan Tourism Development Corporation Ltd., and Pradeep Bohra, General Manager, Palace on Wheels

Railway Board
Rail Bhavan, 1, Raisina Road,
New Delhi - 110001, India
www.indianrailways.gov.in
With special thanks to Seema Sharma, Director, Information & Publicity, Railway Board, and Siddharth Singh, Deputy Director Public Relations, Railway Board

Railroad Museum of Pennsylvania, PHMC
P.O. Box 15,
Strasburg, PA17579
www.rrmuseumpa.org
With special thanks to Dodie Robbins, Nicholas Zmijewski, Charles Fox, and Deborah Reddig

Rewari Steam Loco Shed
Northern Railways, Rewari, Haryana - 123110, India
www.rewaristeamloco.com
With special thanks to Shyam Bihari Gautam, Sr. Section Engineer, Rewari Steam Loco Shed

Ribble Steam Railway Museum
Chain Caul Road,
PR2 2PD, UK
www.ribblesteam.org.uk
With special thanks to Howard Fletcher, Terri Hearty, Jayne Waring, and Chris Mills

SCMG Enterprises Limited
The Science Museum,
Exhibition Road,
London, SW7 2DD, UK
With special thanks to Sophia Brothers and Wendy Burford

The Merchant Navy Locomotive Preservation Society Ltd
(Owners of 35028 *Clan Line*)
12 Inglewood Avenue,
Camberly,
Surrey, GU15 1RJ, UK
www.clan-line.org.uk
With special thanks to Mr R.F. Abercrombie, Tim Robbins, Peter Starks, and Alan French

Venice Simplon-Orient-Express Limited
Shackleton House,
4, Battle Bridge Lane,
London, SE1 2HP, UK
www.orient-express.com
With special thanks to Andrew Cook, Victoria Christie, Jeff Monk, Julian Clark, and Pat Thompson

Virginia Museum of Transportation
303 Norfolk Avenue SW,
Roanoke, VA 24016
www.vmt.org
With special thanks to Beverly Fitzpatrick and Fran Ferguson

PICTURE CREDITS AND VEHICLE OWNERS

Key to museums/contributors
B&O Railroad Museum (BORM)
Didcot Railway Centre (DRC)
Eisenbahnfreunde Traditionsbahnbetriebswerk Stassfurt e.V. (ETS)
Ffestiniog & Welsh Highland Railways (FWHR)
Harzer Schmalspurbahner (HSB)
The National Railway Museum, India (NRMI)
The National Railway Museum, York (NRMY)
Railroad Museum of Pennsylvania, PHMC (RMP)
Rewari Steam Loco Shed (RSLS)
Ribble Steam Railway (RSR)
Virginia Museum of Transportation (VMT)

(Key: a-above; b-below/bottom; c-center; f-far; l-left; r-right; t-top)

1 Dorling Kindersley: Gary Ombler / Courtesy of RSR / Science Museum Group. **2-3 Dorling Kindersley:** Gary Ombler / Courtesy of BORM. **4 Dorling Kindersley:** Gary Ombler / Courtesy of NRMY / Science Museum Group. **5 Dorling Kindersley:** Gary Ombler / Courtesy of BORM (bl); Gary Ombler / Courtesy of NRMY / Science Museum Group (br). **6 Dorling Kindersley:** Gary Ombler / Courtesy of Adrian Shooter (bl); Gary Ombler / Courtesy of DRC (br). **7 Dorling Kindersley:** Gary Ombler / Courtesy of FWHR (bl); Gary Ombler / Courtesy of The Merchant Navy Locomotive Preservation Society Ltd (br). **8 Dorling Kindersley:** Gary Ombler / Courtesy of BORM (bl); Gary Ombler / Courtesy of NRMY / Science Museum Group (br). **9 Dorling Kindersley:** Gary Ombler / Courtesy of VMT (bl); Gary Ombler / Courtesy of Hitachi Rail Europe Ltd (br). **10-11 Dorling Kindersley:** Gary Ombler / Courtesy of NRMY / Science Museum Group. **12 Getty Images:** (c). **13 Corbis:** Underwood & Underwood (br). **TopFoto.co.uk:** (ca). **14 Corbis:** Heritage Images (tl). **Science & Society Picture Library:** Science Museum (bl). **14-15 Getty Images:** SSPL / National Rail Museum. **16 Dorling Kindersley:** Mike Dunning / Courtesy of The Science Museum, London (cl). **Science & Society Picture Library:** National Railway Museum (tr). **16-17 Dorling Kindersley:** Mike Dunning / Courtesy of NRMY (tc). **SuperStock:** Science and Society (bc). **17 The Bridgeman Art Library:** Science Museum, London, UK (cb). **Dorling Kindersley:** Mike Dunning / Courtesy of NRMY (br). **Science Museum, London :** (cr). **18 Dorling Kindersley:** Mike Dunning / Courtesy of NRMY (bc, cr). **Getty Images:** Gallo Images (tl). **19 Dorling Kindersley:** Gary Ombler / NRMY / Science & Society Picture Library, London (tc, tr); Gary Ombler / Courtesy of NRMY / Science Museum Group (c). **20-21 Dorling Kindersley:** Gary Ombler / Courtesy of NRMY / Science Museum Group (all). **22-23 Getty Images:** SSPL / NRM / Pictorial Collection (c). **24 Getty Images:** SSPL / NRM / Pictorial Collection (cra). **Milepost:** (tc). **Smithsonian Institution, Washington,**

DC, USA: (bc, cl). **25 Alamy Images:** The Art Gallery (cb). **colour-rail.com:** (tc, br). **Milepost:** Orion Books / Blandford / Clifford & Wendy Meadway (ca, clb). **26-27 Corbis:** Michael Nicholson (c). **26 Corbis:** Hulton-Deutsch Collection (tl); Michael Nicholson (ftl). **Getty Images:** SSPL / NRM / Pictorial Collection (bl). **28 The Bridgeman Art Library:** Peter Newark American Pictures (cr); National Railway Museum, York, Uk (tr). **Dorling Kindersley:** Gary Ombler / Courtesy of RMP (b). **29 Baltimore and Ohio Railroad:** (bc). **Dorling Kindersley:** Gary Ombler / Courtesy of BORM (cl). **Mary Evans Picture Library:** (tr). **TopFoto.co.uk:** ullsteinbild (crb). **30 Dorling Kindersley:** Gary Ombler / Courtesy of BORM (cl, b). **31 Baltimore and Ohio Railroad:** (tr). **Dorling Kindersley:** Gary Ombler / Courtesy of BORM (tl, bl, br). **Masterfile:** (clb). **Wikipedia:** Urmelbeauftragter (c). **32-33 Dorling Kindersley:** Gary Ombler / Courtesy of BORM. **34 Corbis:** Philip Gendreau / Bettmann (c). **35 Getty Images:** De Agostini (br). **36 Corbis:** Bettmann (br). **Dorling Kindersley:** Gary Ombler / Courtesy of BORM (cl, bl). **36-37 Dorling Kindersley:** Gary Ombler / Courtesy of BORM. **37 Dorling Kindersley:** Gary Ombler / Courtesy of BORM (t). **Golden Spike National Historic Site Promontory Summit, Utah :** (bl, br). **38 BORM:** (tl). **Dorling Kindersley:** Gary Ombler / Courtesy of BORM (c, cr, clb, b). **39 Dorling Kindersley:** Gary Ombler / Courtesy of BORM. **40-41 Dorling Kindersley:** Gary Ombler / Courtesy of BORM (all). **42 Corbis:** Steve Crise / Transtock (tl); David Pollack (c). **43 Alamy Images:** Niall McDiarmid (tl); Visions of Americak LLC (br). **Corbis:** Bettmann (crb, cr); (cra). **Getty Images:** (tr). **44 Mary Evans Picture Library:** (cla). **44-45 Dorling Kindersley:** Gary Ombler / Courtesy of FWHR (bl); Mike Dunning / Courtesy of NRMY (bc). **45 The Bridgeman Art Library:** Ironbridge Gorge Museum, Telford, Shropshire, UK (tr). **Brian Stephenson/RAS:** (ca). **Dorling Kindersley:** Gary Ombler / Courtesy of NRMY / Science Museum Group (cl); Gary Ombler / Courtesy of DRC (cr). **46 Dorling Kindersley:** Gary Ombler / Courtesy of the Verkehrshaus der Schweiz, Luzern, Switzerland (c). **Tobias Koehler:** (tr). **Alex Leroy:** (t). **46-47 Dorling Kindersley:** Gary Ombler (bc). **47 Alamy Images:** Didier Zylberyng (tc). **Getty Images:** SSPL (cra). **Verkehrsmuseum Dresden:** (cl). **48 Alamy Images:** Pictorial Press Ltd (bl). **Getty Images:** SSPL / Science Museum (tl). **48-49 Corbis:** Hulton-Deutsch Collection (c). **50 Alamy Images:** ImagesEurope (cb). **Didcot Railway Centre:** (cl). **50-51 Steam Picture Library:** (tc). **51 Dorling Kindersley:** Gary Ombler / Courtesy of DRC (bl). **Mary Evans Picture Library:** (ca). **Science & Society Picture Library:** NRM / Pictorial Collection (br). **Steam Picture Library:** (clb). **52 Dorling Kindersley:** Mike Dunning / Courtesy of NRMY (bc). **Getty Images:** (cla). **52-53 Dorling Kindersley:** Mike Dunning / Courtesy of NRMY (bc). **Science & Society Picture Library:** NRM / Pictorial Collection (tc). **53 Dorling Kindersley:** Gary Ombler / Courtesy of DRC (c); Gary Ombler / Courtesy of NRMY / Science Museum Group (b). **54-55 Alamy Images:** The Keasbury-Gordon Photograph Archive (c). **56 akg-images:** (tr). **NSW Government State Records:** (cla). **57

Dorling Kindersley:** Deepak Aggarwal / Courtesy of NRMI (c). **Mary Evans Picture Library:** (br). **Danie van der Merwe:** (tl). **58-59 Dorling Kindersley:** Gary Ombler / Courtesy of Adrian Shooter. **60 Science & Society Picture Library:** NRM / Pictorial Collection (c). **61 Alamy Images:** North Wind Picture Archives (ca). **The Bridgeman Art Library:** British Library, London, UK (br). **62-63 colour-rail.com:** (tc). **Dorling Kindersley:** Mike Dunning / Courtesy of NRMY (bc). **62 Brian Stephenson/RAS:** (cla). **John Whiteley:** (crb). **63 Edward Gately:** (c). **64 colour-rail.com:** (tr). **Steam Picture Library:** (cl). **Brian Stephenson/RAS:** (c, clb, bc). **65 colour-rail.com:** (clb, br). **Getty Images:** SSPL / National Railway Museum (cra). **Brian Stephenson/RAS:** (cr). **TfL from the London Transport Museum collection :** (tc, bc). **66-67 Getty Images:** SSPL. **68 Dorling Kindersley:** Gary Ombler / Courtesy of RMP (c, cr, clb, b). **Railroad Museum of Pennsylvania:** (tl). **69 Dorling Kindersley:** Gary Ombler / Courtesy of RMP. **70-71 Dorling Kindersley:** Gary Ombler / Courtesy of RMP (all). **72-73 akg-images:** Universal Images Group. **74 4Corners:** Damm Stefan (c). **Getty Images:** (tr). **75 Canadian Pacific Railway:** Canadian Pacific Archives NS.1960a (cr); Canadian Pacific Archives NS.12756 (crb). **Corbis:** Wayne Barrett & Anne McKay / All Canada Photos (tl); Hulton-Deutsch Collection (cra); Sean Sexton Collection (tr). **Glenbow Museum:** (br). **76 Dorling Kindersley:** Gary Ombler / Courtesy of the Verkehrshaus der Schweiz, Luzern, Switzerland (c); Gary Ombler / Courtesy of FWHR (bl). **Milepost:** (crb). **77 Brian Stephenson/RAS:** (tr). **David Wilcock:** (bc). **Dorling Kindersley:** Gary Ombler / Courtesy of NRMY / Science Museum Group (tl); Gary Ombler / Courtesy of FWHR (c, cb). **78 Dorling Kindersley:** Gary Ombler / Courtesy of FWHR (c, cr, clb, b). **Ffestiniog & Welsh Highland Railways:** Andrew Thomas (tl). **79 Dorling Kindersley:** Gary Ombler / Courtesy of FWHR. **80-81 Dorling Kindersley:** Gary Ombler / Courtesy of FWHR (all). **82 Dorling Kindersley:** Deepak Aggarwal / Courtesy of NRMI (cl, bl). **82-83 Dorling Kindersley:** Gary Ombler / Courtesy of Adrian Shooter (t); Gary Ombler / Courtesy of RMP (c). **83 colour-rail.com:** (cr). **Dorling Kindersley:** Gary Ombler / Courtesy of FWHR (ca). **Keith Fender:** (cl). **Milepost:** (br). **84-89 Dorling Kindersley:** Gary Ombler / Courtesy of Adrian Shooter (all). **90-91 Getty Images:** Imagno / Hulton Archive (c). **92-93 Dorling Kindersley:** Gary Ombler / Courtesy of NRMY / Science Museum Group. **94 Mary Evans Picture Library:** (br). **95 akg-images:** (br). **Mary Evans Picture Library:** (ca). **96 colour-rail.com:** (tc, bl). **Brian Stephenson/RAS:** (c). **97 Milepost:** (cra, tc, bc). **Brian Stephenson/RAS:** (bl). **98 TfL from the London Transport Museum collection :** (tr). **David Wilcock:** (cla). **Dorling Kindersley:** Gary Ombler / Courtesy of DRC (bc). **98-99 Dorling Kindersley:** Gary Ombler / Courtesy of NRMY / Science Museum Group (b). **99 Milepost:** (tr). **Brian Stephenson/RAS:** (cr). **David Wilcock:** (ca). **Dorling Kindersley:** Gary Ombler / Courtesy of DRC (bl, br). **100 Didcot

Railway Centre:** (tl). **Dorling Kindersley:** Gary Ombler / Courtesy of DRC (c, cr, bc). **101 Dorling Kindersley:** Gary Ombler / Courtesy of DRC (c). **102-103 Dorling Kindersley:** Gary Ombler / Courtesy of DRC (all). **104 Dorling Kindersley:** Gary Ombler / Courtesy of the DB Museum, Nurnburg, Germany (br); Gary Ombler / Courtesy of the Musee de Chemin de Fer, Mulhouse (tc). **Getty Images:** ND / Roger Viollet (tr). **Brian Stephenson/ RAS:** (cla). **104-105 Dorling Kindersley:** Gary Ombler / Courtesy of the Musee de Chemin de Fer, Mulhouse (c). **Brian Stephenson/RAS:** (br, tr). **106 Mary Evans Picture Library:** Epic / Tallandier (tl); (bl). **106-107 Mary Evans Picture Library:** (c). **108 Railroad Museum of Pennsylvania:** (tl). **Dorling Kindersley:** Gary Ombler / Courtesy of RMP (c, cr, clb, b). **109-111 Dorling Kindersley:** Gary Ombler / Courtesy of RMP (all). **112 Milepost:** (tc). **Brian Stephenson/RAS:** (cl). **113 Dorling Kindersley:** Gary Ombler / Courtesy of VMT (tr). **Alamy Images:** John Wingfield (cb). **Milepost:** (bc). **114 Virginia Museum of Transportation:** (tl). **Dorling Kindersley:** Gary Ombler / Courtesy of VMT (c, cr, clb, b). **115-117 Dorling Kindersley:** Gary Ombler / Courtesy of VMT (all). **118-119 Corbis:** J. S. Johnston (c). **120 Dorling Kindersley:** Deepak Aggarwal / Courtesy of NRMI (cl); Gary Ombler / Courtesy of HSB (c). **Peter Johnson:** (cr). **Milepost:** (bc). **121 Brian Stephenson/RAS:** (tr). **Dorling Kindersley:** Deepak Aggarwal / Courtesy of NRMI (cl); Gary Ombler / Courtesy of FWHR (cr); Gary Ombler / Courtesy of RMP (br). **122 Corbis:** Douglas Kirkland (clb). **The Library of Congress, Washington DC:** (tr). **www.56thparallel.com/:** (cl). **123 Alamy Images:** Horizons WWP (cl); Andrey Semenov (crb). **Corbis:** Wolfgang Kaehler (cl). **Getty Images:** UIG (tr, ca). **Mary Evans Picture Library:** Illustrated London News (cr); Imagno (br, cra). **124 Stephen Middleton:** (cl). **Siemens AG, Munich/Berlin:** (tr, cb). **124-125 Dorling Kindersley:** Mike Dunning / Courtesy of NRMY (bc). **125 Alamy Images:** David Askham (cr). **Dorling Kindersley:** Gary Ombler / Courtesy of BORM (tl, tr). **126-127 Dorling Kindersley:** Gary Ombler / Courtesy of DRC. **128 Corbis:** David Pollack (c). **129 Dorling Kindersley:** Baltimore and Ohio Railroad (ca). **Getty Images:** SSPL (br). **130 Brian Stephenson/RAS:** A.W. Croughton (cl, tc). **David Wilcock:** (bl). **130-131 Dorling Kindersley:** Gary Ombler / Courtesy of HSB. **131 Milepost:** (cra). **Brian Stephenson/ RAS:** C.R.L. Coles (br); (tc, cr). **132-133 Getty Images:** UIG (c). **132 Archives New Zealand:** (bl). **134 Dorling Kindersley:** Gary Ombler / Courtesy of RMP (bl). **Getty Images:** SSPL / National Railway Museum (br). **Milepost:** (tr). **134-135 Dorling Kindersley:** Gary Ombler / Courtesy of DRC. **135 Brian Stephenson/RAS:** F.R. Hebron (tl); T.G. Hepburn (tr). **Dorling Kindersley:** Gary Ombler / Courtesy of DRC (cr). **136 Didcot Railway Centre:** Bill Turner (tl). **Dorling Kindersley:** Gary Ombler / Courtesy of DRC (c, cr, clb, b). **137-139 Dorling Kindersley:** Gary Ombler / Courtesy of DRC (all). **140 Corbis:** Chris Hellier (cr); Claude Salhani / Sygma (ca). **Getty Images:**

Print Collector (bl). **141 Alamy Images:** imageBROKER (br); peter jordan (cra). **Corbis:** Wolfgang Kaehler (tr); Rob Tilley (cr). **Getty Images:** E+ (crb). **142 Alexander Turnbull Library, National Library Of New Zealand, Te Puna Matauranga o Aotearoa:** (cl). **colour-rail.com:** (cl). **Brian Stephenson/RAS:** (tr). **142-143 Dorling Kindersley:** Gary Ombler / Courtesy of DRC (b). **143 Didcot Railway Centre:** Frank Dumbleton (c). **Dorling Kindersley:** Gary Ombler / Courtesy of DRC (cra). **Brian Stephenson/RAS:** (cla). **144 colour-rail.com:** (cla). **Dorling Kindersley:** Gary Ombler / Courtesy of RMP (tr, cra); Gary Ombler / Courtesy of RSR / Science Museum Group (bl). **144-145 Dorling Kindersley:** Gary Ombler / Courtesy of FWHR (b). **145 David Wilcock:** (c). **Dorling Kindersley:** Gary Ombler / Courtesy of DRC (tl, tr); Deepak Aggarwal / Courtesy of NRMI (br). **146 Chris Doering:** (cr). **Dorling Kindersley:** Gary Ombler / Courtesy of RMP (tr); Deepak Aggarwal / Courtesy of RSLS (cla); Gary Ombler / Courtesy of DRC (bc). **146-147 Dorling Kindersley:** Gary Ombler / Courtesy of ETS (c); Gary Ombler / Courtesy of RMP (b). **147 Kevin Andrusia:** (tl). **colour-rail.com:** (cra). **Dorling Kindersley:** Gary Ombler / Courtesy of DRC (cl).Gary Ombler / Courtesy of RMP (c). **148 Getty Images:** SSPL / National Railway Museum (tl); SSPL / NRM / Pictorial Collection (bl). **148-149 Getty Images:** SSPL / National Railway Museum (c). **150 Brian Stephenson/RAS:** (tr, clb). **150-151 Science & Society Picture Library:** National Railway Museum (c). **151 colour-rail.com:** (tl). **Corbis:** Hulton-Deutsch Collection (br). **Dorling Kindersley:** Gary Ombler / Courtesy of BORM (tr). **Garn Collection:** Borsig (clb). **152 colour-rail.com:** (tl). **Dorling Kindersley:** Gary Ombler / Courtesy of NRMY / Science & Society Picture Library, London (c). **Science & Society Picture Library:** National Railway Museum (cr). **153 Alamy Images:** i4images rm (tr). **Dorling Kindersley:** Gary Ombler / Courtesy of NRMY / Science Museum Group (tl). **154-155 Dorling Kindersley:** Gary Ombler / Courtesy of NRMY / Science Museum Group (all). **156 colour-rail.com:** (ca). **Milepost:** (bc). **156-157 colour-rail.com. 157 Canada Science &Technology Museum:** (tl). **Dorling Kindersley:** Gary Ombler / Courtesy of BORM (cra); Gary Ombler / Courtesy of RMP (crb). **PROV:** (tr). **The Library of Congress, Washington DC:** (br). **158 akg-images:** ullstein bild / ullstein - Jaffø (cla). **Dorling Kindersley:** Gary Ombler / Courtesy of the Verkehrshaus der Schweiz, Luzern, Switzerland (tr). **Mary Evans Picture Library:** Sueddeutsche Zeitung Photo (bc, br). **Steam Picture Library:** (cb). **158-159 Dorling Kindersley:** Gary Ombler / Courtesy of the Musee de Chemin de Fer, Mulhouse (c). **159 Corbis:** Bettmann (bl). **Dorling Kindersley:** Gary Ombler / Courtesy of RMP (tr). **Brian Stephenson/RAS:** (br). **160 Dorling Kindersley:** Deepak Aggarwal / Courtesy of NRMI (c); Gary Ombler / Courtesy of ETS (crb); Gary Ombler / Courtesy of HSB (br). **160-161 Dorling Kindersley:** Gary Ombler / Courtesy of the Verkehrshaus der Schweiz, Luzern, Switzerland. **161 Dorling Kindersley:** Gary Ombler / Courtesy of RMP (tr); Gary Ombler / Courtesy of BORM (br); Gary Ombler / Courtesy of HSB (cr). **Keith Fender:** (bl). **Brian Stephenson/RAS:** T.G. Hepburn (cb).

162 Railroad Museum of Pennsylvania: (tl). **Dorling Kindersley:** Gary Ombler / Courtesy of RMP (c, cr, b). **163-165 Dorling Kindersley:** Gary Ombler / Courtesy of RMP (all). **166-167 Dorling Kindersley:** Gary Ombler / Courtesy of ETS. **168 Corbis:** Swim Ink 2, LLC / Fred Chance (c). **169 Corbis:** Bettmann (tr); PoodlesRock (ca). **170 colour-rail.com:** (cla). **Brian Stephenson/RAS:** (cl, cb). **Wikipedia:** Hans-Peter Scholz (bl). **170-171 Alamy Images:** jozef sedmak (bc). **Dorling Kindersley:** Gary Ombler / Courtesy of ETS (tc). **171 colour-rail.com:** (cr). **Dorling Kindersley:** Gary Ombler / Courtesy of BORM (tr); Deepak Aggarwal / Courtesy of RSLS (br). **Getty Images:** SSPL (cla). **172 Brian Stephenson/RAS:** (tl). **Dorling Kindersley:** Gary Ombler / Courtesy of ETS (c, cr, clb, b). **173-175 Dorling Kindersley:** Gary Ombler / Courtesy of ETS (all). **176-177 Corbis:** Hulton-Deutsch Collection (c). **178 Dorling Kindersley:** Gary Ombler / Courtesy of VMT (tl, cl). **178-179 Dorling Kindersley:** Gary Ombler / Courtesy of RMP (b). **Ted Ellis:** (tc). **179 Dorling Kindersley:** Gary Ombler / Courtesy of BORM (tr); Gary Ombler / Courtesy of VMT (cr). **180 Dorling Kindersley:** Gary Ombler / Courtesy of RMP (cl); Gary Ombler / Courtesy of VMT (bl). **180-181 Dorling Kindersley:** Gary Ombler / Courtesy of BORM (t, cl); Gary Ombler / Courtesy of VMT (b). **182 Courtesy of Norfolk & Western Historical Society (NWHS Collection):** (tl). **Dorling Kindersley:** Gary Ombler / Courtesy of VMT (c, cr, clb, b). **183-185 Dorling Kindersley:** Gary Ombler / Courtesy of VMT (all). **186 Dorling Kindersley:** Gary Ombler / Courtesy of DRC (cl); Gary Ombler / Courtesy of RSR / Science Museum Group (bl). **186-187 Dorling Kindersley:** Gary Ombler / Courtesy of DRC (t); Gary Ombler / Courtesy of RSR / Science Museum Group (c). **187 colour-rail.com:** (cra). **Keith Fender:** (br). **Brian Stephenson/RAS:** (bl, cb). **188 colour-rail.com:** (tl). **Dorling Kindersley:** Gary Ombler / Courtesy of RSR / Science Museum Group (c, cr, clb, b). **189-191 Dorling Kindersley:** Gary Ombler / Courtesy of RSR / Science Museum Group (all). **192 colour-rail.com:** (bc). **Milepost:** (clb). **192-193 Marek Ciesielski:** (bc). **Keith Fender:** (tc). **193 colour-rail.com:** (cra). **Dorling Kindersley:** Gary Ombler / Courtesy of the DB Museum, Nurnburg, Germany (tr); Gary Ombler / Courtesy of ETS (br). **Keith Fender:** (clb). **194 The Blue Train:** (tr). **Corbis:** Prisma Bildagentur AG (bl). **195 Alamy Images:** Art Directors & TRIP (tr); Nioreon (crb); Gabbro (fcra); Pete Titmuss (cra); Stock Connection Blue (fcrb). **Corbis:** Eric Nathan / Loop Images (cr); Brian A. Vikander (br). **196 Dorling Kindersley:** Gary Ombler / Courtesy of the Musee de Chemin de Fer, Mulhouse. **Keith Fender:** (cla). **Brian Stephenson/RAS:** (tr). **197 colour-rail.com:** (tr, cr). **Brian Stephenson/RAS:** (br). **198 Brian Stephenson/RAS:** (tr, crb, bl). **Dorling Kindersley:** Gary Ombler / Courtesy of RSR / Science Museum Group (cl). **199 colour-rail.com:** (tc, cla). **Milepost:** (bc). **Brian Stephenson/RAS:** (cra). **200 Virginia Museum of Transportation:** (tl). **Dorling Kindersley:** Gary Ombler / Courtesy of VMT (c, cr, clb, b). **201-203 Dorling Kindersley:** Gary Ombler / Courtesy of VMT (all). **204 Dorling Kindersley:** Gary Ombler / Courtesy of VMT (tr, crb); Deepak Aggarwal /

Courtesy of RSLS (tc). **Milepost:** (c). **205 Dorling Kindersley:** Deepak Aggarwal / Courtesy of RSLS (tl, c). **Milepost:** (crb). **206 Aditya Kaushal: https://www.flickr.com/photos/45613074@N06/:** (tl). **Dorling Kindersley:** Deepak Aggarwal / Courtesy of RSLS (c, cr, clb, b). **207-209 Dorling Kindersley:** Deepak Aggarwal / Courtesy of RSLS (all). **210 Milepost:** (cra). **Brian Stephenson/RAS:** (cla, cb). **211 Brian Stephenson/RAS:** (tr). **Dorling Kindersley:** Gary Ombler / Courtesy of HSB (ca); Gary Ombler / Courtesy of FWHR (c). **212 Ffestiniog & Welsh Highland Railways:** Andrew Thomas (tl). **Dorling Kindersley:** Gary Ombler / Courtesy of FWHR (c, cr, b). **213-215 Dorling Kindersley:** Gary Ombler / Courtesy of FWHR (all). **216 Dorling Kindersley:** Gary Ombler / Courtesy of DRC (tr, clb); Gary Ombler / Courtesy of VMT (cla); Gary Ombler / Courtesy of RMP (bc). **216-217 Dorling Kindersley:** Gary Ombler / Courtesy of ETS. **217 Garn Collection:** (tc). **Dorling Kindersley:** Gary Ombler / Courtesy of VMT (cla); Gary Ombler / Courtesy of DRC (r, cr); Gary Ombler / Courtesy of BORM (br). **218-219 Dorling Kindersley:** Gary Ombler / Courtesy of NRMY / Science Museum Group. **220 Getty Images:** SSPL (c). **221 Alamy Images:** Everett Collection Historical (bc). **Getty Images:** Sankei Archive (cr). **222 Dorling Kindersley:** Gary Ombler / Courtesy of ETS (cl). **Keith Fender:** (c, cra, tc). **222-223 Dorling Kindersley:** Gary Ombler / Courtesy of the Museum of Transportation, St Louis, Missouri (bc). **223 Dorling Kindersley:** Gary Ombler / Courtesy of BORM (t, br). **Keith Fender:** (cra). **Brian Stephenson/RAS:** (cl). **224 Harzer Schmalspurbahnen:** (tl). **Dorling Kindersley:** Gary Ombler / Courtesy of HSB (c, cr, b). **225-227 Dorling Kindersley:** Gary Ombler / Courtesy of HSB (all). **228 Amtrak:** Amtrak History and Archives (bl). **colour-rail.com:** (tr). **Getty Images:** UIG (cl). **228-229 Alamy Images:** Craig Yates T (bc). **229 Keith Fender:** (br). **Roger Lalonde:** (cr). **Brian Stephenson/RAS:** (cra). **230-231 Corbis:** José Fuste Raga (cl). **232 Roger Bastin:** (tl). **Dorling Kindersley:** Gary Ombler / Courtesy of Dampf-Plus GmbH (c, cr, clb, b). **233-235 Dorling Kindersley:** Gary Ombler / Courtesy of Dampf-Plus GmbH (all). **236 Keith Fender:** (cra). **TopFoto.co.uk:** RIA Novosti (cl). **Dorling Kindersley:** Gary Ombler / Courtesy of RSR / Science Museum Group (tr, cla); Gary Ombler / Courtesy of ETS (bl). **237 colour-rail.com:** (cr). **Dorling Kindersley:** Gary Ombler / Courtesy of ETS(tr, b). **TfL from the London Transport Museum collection :** (ca). **238 Courtesy of the James Northfield Heritage Art Trust ©. Amanda Slater:** (cl). **239 Alamy Images:** age fotostock (crb); Eric Nathan (br); Robert Harding Picture Library (cra). **National Railway Museum Australia:** (ftr). **Great Southern Rail:** (cr, cl). **National Archives of Australia:** (tr). **240 Keith Fender:** (tr, br). **Charles P Friel:** (ca). **Mirko Schmidt:** (bl). **240-241 Dorling Kindersley:** Gary Ombler / Courtesy of RMP (c). **241 Keith Fender:** (tl, tr). **Don Oltmann:** (bc). **242-243 Dorling Kindersley:** Gary Ombler / Courtesy of DB Schenker. **244 Michael Rhodes:** (c). **245 Eurostar:** (bc). **Keith Fender:** (cr). **246 Alamy Images:** Robert Harding Picture Library (br). **Dorling

Kindersley:** Gary Ombler / Courtesy of Transrapid, Lathen, Germany (bl). **246-247 colour-rail.com:** (c). **TopFoto.co.uk:** RIA Novosti (tc). **247 Keith Fender:** (tr, bc). **Milepost:** Brian Solomon (cra, cla). **248 Alamy Images:** Malcolm Case-Green (cra); Lordprice Collection (c). **Daniel Minaca:** (bl). **249 Alamy Images:** qaphotos.com (cr, fcr, crb). **Corbis:** Ian Cumming / Design Pics (tr); PictureNet (br). **250-251 colour-rail.com:** Bob Sweet (c). **250 colour-rail.com:** Bob Sweet (c). **Keith Fender:** (tr). **Milepost:** (cl). **251 colour-rail.com:** (tl). **Dorling Kindersley:** Gary Ombler / Courtesy of HSB (cr).**Keith Fender:** (tr, cr). **252 Alamy Images:** Prisma Bildagentur AG (bl). **Dorling Kindersley:** Gary Ombler / Courtesy of ETS (cla). **Keith Fender:** (tr, cr). **252-253 Dorling Kindersley:** Gary Ombler / Courtesy of DB Schenker (b). **253 colour-rail.com:** (tl); Bob Sweet (cla). **Keith Fender:** (crb). **254 Alamy Images:** Caro (c); Indiapicture (cr). **Dorling Kindersley:** Deepak Aggarwal / Courtesy of Safdarjung Railway Station (crb, b). **www.palaceonwheels.net:** (tl). **255-259 Dorling Kindersley:** Deepak Aggarwal / Courtesy of Safdarjung Railway Station (all). **260 Alamy Images:** Gunter Marx (tr). **Keith Fender:** (cl). **Brian Stephenson/RAS:** (cr). **260-261 Dorling Kindersley:** Mike Dunning (bc). **Milepost:** Brian Solomon (c). **261 colour-rail.com:** (tl); Bob Sweet (cra). **Keith Fender:** (cra). **262-263 Dorling Kindersley:** Gary Ombler / Courtesy of Hitachi Rail Europe Ltd. **264 Alamy Images:** Raga Jose Fuste / Prisma Bildagentur AG (c). **265 Imaginechina:** Ren yuming (cr). **With thanks to Venice Simplon-Orient Express Limited www.orientexpress.com:** (bc). **266 Keith Fender:** (cla, bc, tr). **266-267 Craig Walker. 267 Keith Fender:** (br, cb). **Siemens AG, Munich/Berlin:** (tc). **Vossloh AG:** (cr, bl). **268 Alamy Images:** Degas Jean-Pierre / Hemis.fr (tc); Didier Zylberyng (cb); Lemaire Stéphane / Hemis.fr (crb). **Corbis:** Walter Bibikow / JAI (tr); Luke Macgregor / Reuters (bc); Scott S. Warren / National Geographic Society (cl); Arnd Wiegmann / Reuters (cr). **Getty Images:** David Boyer / National Geographic (tl). **268-269 Alamy Images:** Aigars Reinholds (bc). **269 Alamy Images:** David Lyons (tl). **Dorling Kindersley:** Christopher Pillitz (br). **Tony Streeter:** (tr). **270 Dorling Kindersley:** Gary Ombler / Courtesy of The Merchant Navy Locomotive Preservation Society Ltd (c, crb); Sharon Spencer / Courtesy of Venice Simplon Orient Express Ltd (bc). **Peter Starks:** (tl, cra). **271-273 Dorling Kindersley:** Gary Ombler / Courtesy of The Merchant Navy Locomotive Preservation Society Ltd (all). **274-277 Dorling Kindersley:** Gary Ombler / Courtesy of Venice Simplon Orient Express Ltd (all). **278 Keith Fender:** (c, tr). **Imaginechina:** Gao yuwen (crb). **Milepost:** (bl). **279 Alamy Images:** Susan Isakson (tc); Iain Masterton (cr). **Japan National Tourism Organization :** Awajiya (br). **Ilya Semenoff:** (bl). **280 Alamy Images:** Richard Bradley (bc). **Corbis:** Franck Guiziou / Hemis (clb); Topic Photo Agency (bl). **280-281 Alamy Images:** Axel Schmies / Novarc Images (tc); Neil Setchfield (bc). **281 Alamy Images:** Angelo Cavalli (br); Paul Springett 06 (tr); Michele and Tom

Grimm (bc); Eddie Linssen (clb). **Corbis:** Massimo Borchi / Atlantide Phototravel (crb); Wu Hong / EPA (tc); Jon Hicks (ca); Ricky Leaver / Loop Images (cra). **282 Alamy Images:** Avpics (bl). **colour-rail.com:** (c). **Keith Fender:** (tr). **283 Dorling Kindersley:** Gary Ombler / Courtesy of Hitachi Rail Europe Ltd (b). **Keith Fender:** (tc). **Getty Images:** Gamma-Rapho (cb). **Brian Stephenson/RAS. 284 Alamy Images:** Maurice Savage (tl). **Dorling Kindersley:** Gary Ombler / Courtesy of Hitachi Rail Europe Ltd (c, cr, b). **285-289 Dorling Kindersley:** Gary Ombler / Courtesy of Hitachi Rail Europe Ltd (all). **290-291 Alamy Images:** John Kellerman (c). **292 Bombardier Transportation:** (cla, c). **Vossloh AG:** (br). **Chris Wallace:** (tr). **293 Keith Fender:** (cla, cra). **Siemens AG, Munich/Berlin:** (cr, br, bl). **Vossloh AG:** (tr). **294-295 Dorling Kindersley:** Gary Ombler / Courtesy of DRC. **296 Dorling Kindersley:** (bc); Mike Dunning / Courtesy of The Science Museum, London (bl). **297 Dorling Kindersley:** Gary Ombler / Courtesy of RMP (cr/rim brake). **299 Dorling Kindersley:** Mike Dunning / Courtesy of The Science Museum, London (fcl, c, cl). **Getty Images:** falcon0125 (cb). **300-301 Dorling Kindersley:** Gary Ombler / Courtesy of DRC (all). **302 Corbis:** Scott Warren / Aurora Photos (bl). **302-303 Dorling Kindersley:** Mike Dunning / Courtesy of NRMY (c). **303 Dorling Kindersley:** Gary Ombler / Courtesy of NRMY / Science Museum Group (t). **305 Dorling Kindersley:** Gary Ombler / Courtesy of VMT (tr); **Dorling Kindersley:** Gary Ombler / Courtesy of HSB (cr). **307 Alamy Images:** Maurice Savage (tr)

All other images © Dorling Kindersley For further information see: www.dkimages.com

Images on title, contents, and introduction
page 1 Deltic prototype
pages 2-3 B&O Class B No. 147 *Thatcher Perkins*
page 4 Rocket
page 5 B&O Class B No. 147 *Thatcher Perkins* (bl), LNER No. 4468 *Mallard* (br)
page 6 DHR B Class No. 19 (bl), King Class No. 6023 *King Edward II* (br)
page 7 NG G16 Beyer-Garratt No. 138 (bl), Merchant Navy Class No. 35028 *Clan Line* (br)
page 8 B&O Class B No. 147 *Thatcher Perkins* (bl), LNER No. 4468 *Mallard* (br)
page 9 N&W GP9 Class No. 521 (bl), Hitachi Javelin No. 395 017 (br)

Images on chapter opener pages
pages 10-11 1804–1838 *Rocket*
pages 32-33 1839–1869 B&O Class B No. 147 *Thatcher Perkins*
pages 58-59 1870–1894 DHR B Class No. 19
pages 92-93 1895–1913 NER Class X1 No. 66 *Aerolite*
pages 126-127 1914–1939 King Class No. 6023 *King Edward II*
pages 166-167 1940–1959 DR No. 52.8184-5
pages 218-219 1960–1979 JR West Shinkansen Series 0 22-141
pages 242-243 1980–1999 DB Schenker Class 92 92042
pages 262-263 After 2000 Hitachi Javelin No. 395 017
pages 294-295 How Railways Work Radstock North Signal Box at Didcot Railway Centre